AN OUTLINE OF
SCIENTIFIC
WRITING

Preface

Today English is the official language of international conferences, and most of the important publications in science and technology now appear in English. Researchers must read English-language journals and books to keep up with advances in their fields. Over twenty percent of the world's population speaks Chinese, but China is still a developing country and few researchers outside of China will understand a scientific publication written in Chinese. While this may not be the case in the twenty-first century, for now, the researcher who wishes to reach a wide readership must publish in English. Therefore, learning how to write a manuscript in English has become part of the researcher's task.

Writing in English can be difficult even for someone who grew up speaking the language, and even more so for anyone who learns it as a foreign language. English derives from many cultures and is constantly evolving. As a result, its grammatical rules are many and complex. We Asians face an additional challenge not shared by our European counterparts: Most Asian dialects, Chinese and Japanese included, do not belong to the same language family as English. There are grammatical constructs that have no corresponding forms in Chinese or Japanese. The task is not merely to translate words, but to understand and use foreign concepts of syntax as well.

Consider the article *the,* which has no grammatical equivalent in Chinese. Knowing when to use it before a noun is intuitive for an English-speaking writer, but quite tricky for the Chinese. Other examples are prepositions such as *at, in,* and *with*. These also have no equivalents in Chinese, which can make usage especially difficult to master. Then there is the plural form of nouns. The European writer is used to seeing both singular and plural nouns in his or her own language, and can easily deal with these forms in English. In Chinese, however, a noun has the same form for both the singular and plural cases. It requires extra vigilance on the part of the writer to be certain of using the correct form in an English sentence.

Such difficulties are by no means insurmountable. With practice, plus attention to the particular challenges faced by the Asian scholar, any of you should be able to write a scientific paper in English that is concise and lucid as well as grammatically correct, even if your vocabulary and understanding of English usage are limited.

There are already many excellent texts on scientific writing in English. Why, then, would an author whose first language is Chinese write another book on the subject? As an author in English for over forty years, I understand the unique writing challenge that we Asians face. This book discusses the style and convention used in scientific publications and is written on a level that can be understood by researchers who learned English as a second language. Rarely will a dictionary be needed. It assumes, however, that the reader can already write *grammatically correct* English and avoid such mistakes as the following:

INCORRECT	CORRECT
The experimental data included circular dichroism and intrinsic viscosity indicated that tropomyosin was a helix-rich rod.	The experimental data including both circular dichroism and intrinsic viscosity indicated that tropomyosin was a helix-rich rod.

I know of some Chinese professors who rely on coworkers fluent in English to polish their scientific writings. While it is fine to have others review your work, this should not be part of the development of the paper. You must learn how to write concisely and lucidly in a well-organized manner. The following illustrates that a sentence can be grammatically correct but difficult to follow:

Supply us with the necessary inputs of relevant equipment and we shall implement the program and accomplish its objects.

Winston Churchill expressed this same idea concisely and far more elegantly:

Give us the tools and we will finish the job.

It is true that a paper is only as good as the underlying research, and writing effort can improve the quality of a poor study. On the other hand, however fine the research, it is wasted if the paper is not read and understood. If, by attention to style and presentation, your paper communicates more effectively, you will have accomplished your purpose.

Professor Julius H. Comroe Jr., former Director of the Cardiovascular Research Institute at the University of California at San Francisco, firmly believed that a concise and lucid writing style furthered the cause of research. I once showed him a review that I had written for *Advances in Protein Chemistry*. One of the paragraphs began, "Needless to say ..." Dr. Comroe crossed out the paragraph and indeed the entire page, writing in the margin, "If it is needless, why say it?" In 1977 I decided to take Dr. Comroe's course on scientific writing. Initially I regarded "How to Write a Paper" with some skepticism, believing that writing skill was innate and could not be learned. Dr. Comroe convinced me, however, that some style rules on this subject can be quite helpful when writing a manuscript.

This book is divided into five parts. Part I deals with the most typical syntax errors. It discusses choice of words, sentence structure and, briefly, the linkage between sentences in paragraphs. Part II, the major part of this book, discusses how to plan a manuscript. It covers the choice of an informative and attractive title and the composition of an abstract (a mini-paper by itself), followed by the standard format of a scientific manuscript: introduction, materials and methods, results, and discussion. Part III explains how to submit a manuscript to a journal and the process of acceptance/rejection and revision of the paper. Part IV discusses the preparation of a poster and some suggestions on oral presentations. In the appendices are the International Union of Pure and Applied Chemistry recommendations on symbols and terminology for quantities and units, and also some common physical and chemical quantities and standard biochemical abbreviations. I find these to be useful reference materials when writing a biochemistry manuscript. Authors in other fields can compile appendices relevant to their own subjects.

The reader who wishes to study scientific writing in more detail is encouraged to read two recent publications: *Essentials of Writing Biomedical Research Papers* by Mimi Zeiger (McGraw-Hill, 1991), and *A Researcher's Guide to Scientific and Medical Illustrations* by Mary Helen Briscoe (Springer-Verlag, 1990). Both books are well regarded and have been translated into Japanese. Incidentally, both authors were associated with the Cardiovascular Research Institute, as I was. Ms. Zeiger and I took Dr. Comroe's course together and Ms. Briscoe helped me prepare numerous illustrations during my tenure at the Institute.

For general writing, I recommend *The Elements of Style* by William Strunk. This is an easy-to-use, classic text that many English teachers consider indispensable. It is, however, somewhat dated in that it recommends using *he* to represent men and women alike. I prefer using *he or she, his or her,* etc. unless it makes the text too unwieldy.

One final recommendation: Do try to think in English when you write a manuscript in English. Text that is translated from one language into another often sounds awkward. It may be difficult to put your thoughts into English, but you will gain facility with practice. You are also advised not to become overly reliant on this or any other writing guide. In the words of Confucius, "Better to have no books than to trust them completely." There is no substitute for practicing a language and developing an ear for its nuances. Only in this way can you master your own scientific writing in English.

Acknowledgments

I am deeply grateful to the late Professor Julius H. Comroe Jr., founding director of the Cardiovascular Research Institute at the University of California at San Francisco, for introducing me to the study of scientific writing. Sadly, he never found time to put his syllabus of scientific writing into book form. Mrs. Julius Comroe has kindly given me permission to quote many of his notes, which constitute a large portion of this text.

I thank Ms. Mary Helen Briscoe and Ms. Mimi Zeiger, both formerly at the Institute, for their friendship and generous assistance over the years. Ms. Briscoe taught me to plot figures using a computer program. Professor Jon Goerke, Ms. Rolinda Wang, and Mr. Isaac Sato, all at the Institute, were most helpful in showing me how to use some software programs. Thanks are also due Professor Tatsuya Samejima, Dr. Lu-Ping Shen, and Dr. Chuen-Shang C. Wu for reading and commenting on the manuscript. I am grateful to Ms. Tina Jeavons, my editor at World Scientific Publishing Co., for her unflagging enthusiasm and sound advice, without which I would have hesitated to have undertaken this project.

Thanks to my ABC (American-born Chinese) daughters Janet and Frances for their help during the preparation of this book. Janet edited the entire manuscript, adding original material, and raised numerous issues of which I had not been aware. I learned a great deal from her comments. Frances also read the entire text and refined my English. Janet and I disagreed on the "Asian modesty" that was the style of the first draft, which I attribute to our differences in cultural background. Nevertheless, in taking her advice I have moderated what she called the "self-deprecating tone" of the text. Last but not least, I thank my wife Yee-Mui for taking on the tedius chore of proofreading the entire manuscript.

Contents

Part I English Grammar

1 Word Choice

A paper will be more readable if words are used economically. Writing concisely may be contrary to common practice in some countries where, I have been told, authors are paid by the number of words published! Always remember that your goal is to facilitate communication, which is accomplished through *concise* and *lucid* writing in a well-organized manner.

A. DELETE UNINFORMATIVE WORDS AND AVOID REDUNDANCY

Using fewer words to convey a message almost always improves readability. It also requires more effort, as the mathematician Pascal once noted to a friend: "I am writing a longer letter than usual, because there is not enough time to write a short one." The examples in the left-hand column below are unnecessarily wordy. The right-hand column displays improved versions. (This side-by-side format for displaying "before and after" examples is used throughout the book.)

brief in duration	brief
sufficient in number	sufficient
The wound was of a serious nature.	The wound was serious.
The solution was red in color.	The solution was red. (Red is understood to be a color.)

Some material in this chapter is from the lecture notes of Julius H. Comroe Jr., Cardiovascular Research Institute, University of California at San Francisco; excerpted by kind permission at Mrs. Julius H. Comroe.

It was precooled before use. It was precooled. (The prefix *pre-* implies *before*.

We repeated the experiment again. We repeated the experiment.

EXERCISES.

1. Rewrite each phrase to eliminate unnecessary and redundant words (Answers are given at the end of the chapter.)
a. carefully investigate
b. past history
c. mix together
d. human volunteers
e. original source
f. advance planning
g. globular in shape
h. more preferable than
i. seem to appear
j. for a period of several minutes
k. The role of cobra toxin is still *a matter for speculation*. (Hint: replace the words in italics.)
l. The work will be completed *in the not-too-distant future*.
m. The reaction rate *was examined and found to vary* considerably.
n. The results *would seem to indicate* the possibility that impurities might be present.
o. *As a result of these experiments it became quite evident that overheating of the samples had occurred.*
p. As far as my own experiments are concerned, they show ...
q. It has been found that ...
r. It is interesting to note that ...
s. Needless to say, ...

B. USE ONE WORD TO REPLACE A PHRASE

Many popular expressions can be expressed as a single word, or are better omitted altogether.

at this point in time	now
the reason was because	because

EXERCISES.

2. Rewrite or eliminate the following phrases.
a. in view of the fact that
b. was observed to be
c. in the near future
d. in most cases
e. it would appear that
f. is suggestive of
g. as to whether
h. in the vicinity of
i. it was evident that
j. in the event that

C. AVOID GRANDILOQUENCE

The word *grandiloquence* is itself grandiose. It implies a pompous style that impresses no one and provokes ridicule. Recall the elegant and memorable words of Sir Winston Churchill: "Give us the tools and we will finish the job." The greatest speeches in history were simple and to-the-point. Abraham Lincoln's Gettysburg address was so brief that his audience was disappointed, but it has since come to be regarded as one of the most effective speeches ever delivered.

The same brevity and clarity should apply to scientific writing. On the left, below, is a sentence in which clarity is obliterated by grandiloquence; on the right, a much-improved version.

The validity of the structural information obtained will depend on the fidelity of reconstitution of the proteolipid in a native conformation under the condition or incorporation into vesicles.	Such structural information is valid only when the reconstituted proteolipid has the native conformation.

EXERCISES.

3. Rewrite the following sentences into concise and effective English.
a. Computations were conducted on the data.
b. It may seem reasonable to suggest that the necrotic effect may possibly be due to toxins.
c. In studies pertaining to identification of phenolic derivatives, drying of the paper gives less satisfactory visualization.
d. A method, which was found to be expedient and not very difficult to accomplish and which possessed a high degree of accuracy on its results, was devised whereby ...

D. AVOID CLICHÉS AND EUPHEMISMS

These are rarely helpful and frequently cryptic.

all in all	(delete)
if and when	if

Some common euphemisms are simply awkward. For instance, such evasions as

The patient expired

The patient passed away

The patient succumbed

The patient breathed his last

The patient has gone to his rest

are better expressed as

The patient died.

EXERCISES.

4.   Rewrite the following phrases and sentences.
a.   unless and until
b.   it goes without saying
c.   within the realm of possibility
d.   We sacrificed the animal.
e.   We performed euthanasia on the animal.

The following terms are usually better omitted or rephrased. Answers have not been provided; the reader should be able to suggest suitable answers.

approximately
a majority of
a number of
an order of magnitude faster
are of the same opinion
as a consequence of
as a matter of fact
As seen from our study, it is evident that
based on the fact that
due to the fact that
first of all
for the reason that
has the capability of
in a satisfactory manner
in order to
in terms of
is defined as

it has been reported by Dr. Lee that
it has long been known that
It is clear that much additional work will be required before a complete understanding.
it is worth pointing out in this context that
it may, however, be noted that
owing to the fact that
the question as to whether
there is reason to believe
with the possible exception of

E. USE SYNONYMS

A *synonym* is a word that has the same or nearly the same meaning as another word. There are two reasons to use synonyms: First, to avoid monotony from using the same term repeatedly.

> The subject demonstrated a marked sensitivity to the allergen. After receiving the medication, she showed marked improvement. This is a marked medical achievement.

The repeated use of *marked* makes this passage tedious. The text can be improved by substituting different synonyms.

> The subject demonstrated a marked sensitivity to the allergen. After receiving the medication, she showed extraordinary improvement. This is a noteworthy medical achievement.

The second reason for using synonyms to express the *precise* shade of meaning for a specific context. *Extraordinary* and *noteworthy* communicate a stronger sense of enthusiasm than *marked*. Goglum [*Cancer Res.* (1945) **5**, 247] has given twenty-two alternatives for the word *marked:*

appreciable	extreme	signal
considerable	great	significant
conspicuous	intense	striking

decided	large	strong
definite	notable	substantial
distinct	noteworthy	unusual
extensive	noticeable	
extraordinary	pronounced	

To this list Dr. Comroe added

advanced	excessive	important
astonishing	goodly	obvious
big	huge	profound
enormous	immense	remarkable

Synonyms for common words can be found in a *thesaurus,* a dictionary, and some word processing programs. Understanding the nuances of synonyms can admittedly be difficult for those of us with English as a second language. The best ways to improve your grasp are to read English-language authors and practice your own writing.

ANSWERS TO EXERCISES.

1. a. investigate (study) b. history c. mix d. volunteers e. source f. planning g. globular h. preferable (to) i. seem j. for several minutes k. speculative l. soon (before long) m. varied n. suggest (were) o. These experiments showed that the samples had evidently been overheated. p. My own experiments show ... q. (delete phrase) r. (delete phrase) s. (delete entire passage)

2. a. because b. was c. soon d. mostly e. (delete) f. suggests g. whether h. near i. evidently j. if (should)

3. a. The data were calculated. b. Necrosis may be caused by toxins. (Toxins may cause necrosis.) c. Phenolic derivatives are easier to see if the paper is left wet. d. An easy, accurate way to ...

4: a. (rewrite the sentence) b. (delete) c. possible d and e. We killed the animal.

2 Sentence Structure

The reader of this book is presumed to be familiar with basic English grammar: parts of speech, tenses, plurals, and so forth. Here we will discuss a few of the more complex rules that sometimes prove troublesome even for those with English as their native language.

A. AGREEMENT OF SUBJECT AND VERB

1. The number of the verb must agree with the number of the subject. A singular subject has a singular verb, and a plural subject a plural verb. It follows that you must correctly identify the subject, which is not necessarily the first or only noun in the sentence.

From this work has come improved antibiotic drugs.	From this work have come improved antibiotic drugs.
An evaluation of the experimental results, as well as the clinical findings, are described.	An evaluation of the experimental results, as well as the clinical findings, is described.

The first example illustrates inverted structure, in which the subject of the sentence, *drugs,* comes after the verb. In the second example, the subject is *evaluation,* not *results* or *findings.*

An easy way to identify the correct subject and verb form is to examine the briefest form of the sentence: Remove prepositional phrases, modifiers,

---

Most of the examples in this chapter are from the lecture notes of JuliusSome material in this chapter is from the lecture notes of Julius H. Comroe Jr., Cardiovascular Research Institute, University of California at San Francisco; excerpted by kind permission at Mrs. Julius H. Comroe.

and so forth, until the sentence is reduced to just its subject, verb, and complement (if any). It usually becomes clear which verb form should be used—especially if the sentence is spoken aloud. Using the preceding example:

An evaluation ... are described.   An evaluation ... is described.

Speaking sentences aloud is a useful check of your writing style. Often the ear will detect what the eye misses, although you cannot always rely on the sound of a sentence, as the next rule shows.

2. Recognize irregular plurals. A common mistake is to use a singular verb with *data, formulae,* and *radii,* which are plural nouns (their singular forms are *datum, formula,* and *radius,* respectively). The error results from failure to recognize irregular plurals, i.e., plural forms that do not end with an *s.*

This data is significant.   These data are significant. (Also note the use of the plural article *these.*

3. When singular and plural subjects are joined by either ... or and neither ... nor, the verb must agree with the nearest subject.

Neither the physical behavior of   Neither the physical behavior of
these substances nor their half-   these substances nor their half-
life data was known.   life data were known.

Either the samples or the   Either the samples or the
apparatus were contaminated.   apparatus was contaminated.

Do not confuse *either ... or* and *neither ... nor* with *either* and *neither,* both of which always take a singular verb.

Either of the explanations is acceptable.

Neither of the samples is large.

A compound sentence with more than one dependent clause must include all verbs, unless they are the same in each clause. That is, if one clause contains a singular subject and another clause a plural subject, both the singular and plural verbs are required.

The tissue was minced and samples incubated.	The tissue was minced and samples were incubated.

All auxiliaries verbs (such as *to be* and *have*) must likewise be included, unless they are the same.

Blood samples have been drawn and measurements have been taken.	Blood samples have been drawn and measurements taken.

Dissimilar auxiliaries cannot be omitted, however.

Blood samples have been drawn and urine collected.	Blood samples have been drawn and urine has been collected.

B. PRONOUN REFERENCE

An *antecedent* is the word, phrase, or clause to which a pronoun refers. A sentence may be confusing if the pronoun and its antecedent are not clearly identifiable. A missing antecedent cannot be assumed to be "obvious from context," and an ambiguous reference should always be corrected. In the statement

The monkey was operated on by the surgeon when he was six weeks old

the reader cannot tell whether it was the surgeon or the monkey who was six weeks old at the time of the surgery. The ambiguity is removed by positioning the pronoun closer to its antecedent.

> The monkey, when he was six weeks old, was operated on by the surgeon.

Better still is to move the relative clause to the beginning, where it will not separate the subject from the principal verb.

> When he was six weeks old, the monkey was operated on by the surgeon.

If the ambiguity cannot be removed by rearranging words, the entire sentence should be rewritten.

Sera were prepared by immunizing rabbits and drying and using them in powdered form for blood grouping.	Sera from immunized rabbits were dried, powdered, and used for blood grouping.

The left-hand version evokes unfortunate images of rabbits being dried and powdered, while the revised version conveys the intended meaning.

C. ACTIVE AND PASSIVE VOICE

English verbs have two voices: *active* and *passive*. In the active voice, the subject performs the action, while in the passive voice, the subject receives the action.

> Frances always wins the award. [Active voice]

> The award is always won by Frances. [Passive voice]

Note the change of the verb from *win* for the active voice to *is won* for the passive. The passive voice always combines some form of the verb *to be* with a past participle.

As fashions change with time, so does the style of scientific writing. Prior to 1900, scientists routinely used the active voice and personal pronouns in their reports, making such statements as, "I made the following experiment," "I cannot say," "I have seen," and "I would point out, however, that ...." The passive voice gradually gained popularity, perhaps from a belief that its impersonal style denoted greater professionalism. *"The consistent overuse and misuse of the passive voice devitalized scientific writing. It became torpid, evasive, and dull, qualities that the writers mistakenly equated with dispassionate objectivity .... Today, the trend is once again toward clarity of expression and the freer, more concise writing that results from habitual use of the active voice"* (Comroe). Style experts now prefer the active voice, which is more direct, sounds more natural, and usually saves words. It clearly identifies who performs the action, and does not necessarily require the use of pronouns *I, we, she,* and so forth.

It was suggested by Dr. Smith that the test be postponed.	Dr. Smith suggested postponing the test.
In these experiments exercise was carried out by the subjects.	In these experiments the subjects exercised.
A detailed description of the apparatus is presented in this Report.	This report presents a detailed description of the apparatus.

This is not to say that you must entirely avoid using the passive voice, which can be quite effective if used sparingly. By placing the receiver of the action as the subject of the sentence, it receives subtle emphasis.

The relationship $F = ma$ was discovered by Newton.

Newton discovered the relationship $F = ma$.

The first version would be appropriate in a text on the history of physics, while the second version could be used in a biography of Sir Isaac Newton.

The passive voice is also used to avoid mentioning the performer of the action when the performer is unimportant, indefinite, unknown, or obvious from context.

Bovine serum albumin was purchased from Sigma.

Eighteen minutes of the tape had been erased.

In the first example, the passive voice is used to focus more on the material than on the purchaser. In the second example, by not mentioning the person who erased the tape, the writer focuses on the erasure—and perhaps avoids a lawsuit as well!

Dr. Comroe illustrated the economy and clarity of the active voice in the following passage, presented first in the passive voice, then in the active voice.

PASSIVE VOICE.

In early studies of longitudinal sections of cardiac muscle by light microscopy, a pattern of cross striations was observed (1–5). The striations were noted to be similar to those seen in skeletal muscle, except that at irregular intervals along the length of the fibers the thin Z-lines were replaced by thick transverse bands. According to Witte (6), the thick bands, or intercalated discs, could not be detected in cardiac muscle in embryonic or early fetal life; the discs were concluded to be of physiologic importance, although precisely what role was played by these discs was conjectural. Later, electron microscopic studies of thin sections of cardiac muscle were performed (7–14). At each intercalated disc there were seen to be present two membranes and a narrow interspace traversing the muscle fibers. The cleft between the transverse membrane was clearly seen to be part of the extracellular space. Thus, it was established that cardiac muscle consists of individual cells with the intercalated discs being sites of cell-to-cell attachment.

ACTIVE VOICE.

Early studies of longitudinal sections of cardiac muscle by light microscopy showed a pattern of cross striations (1-5). The striations resembled those of skeletal muscle, except that at irregular intervals along the fibers, thick transverse bands replaced the thin Z lines. Witte (6) was unable to detect these thick bands, or intercalated discs, in cardiac muscle in embryonic or early fetal life. [He concluded that] the discs were of physiologic importance, but [that] their precise role was conjectural. Later, studies of thin sections of cardiac muscle by electron microscopy (7-14) showed two membranes at each intercalated disc and a narrow interspace traversing the muscle fibers. The cleft between the transverse membranes was clearly part of the extracellular space. [These studies established that] cardiac muscle consists of individual cells and [that] the intercalated discs are sites of cell-to-cell attachment.

Changing the voice from passive to active reduces 165 words to 137. Further, the bracketed words can be deleted without materially affecting the content, a total saving of 22%.

D. NOUNS FROM VERBS

Verbs can express action. For many action verbs there are nouns of similar derivation that expresses the result of the action, for example, *examine/examination* and *perform/performance*. Using the noun form expresses the action indirectly. Your writing will be more vigorous if such nouns are replaced by the verb forms.

By analysis of the data	By analyzing the data
An evaluation of the data was done.	The data were evaluated.
The installation of the new equipment has been carried out.	The new equipment has been installed.

EXERCISES.

Rewrite the following sentences to use the active voice and eliminate unnecessary words (answers follow).
1. He made an examination of the patient.
2. His performance of the tests was adequate.
3. These agents exert their action by inhibition of synthesis of cholesterol by the liver.
4. We made at least two analyses on each specimen.

5. Our preliminary report included a description of the techniques used for the infusion of fluids into the cerebral ventricles of rats.
6. Evaporation of ethanol from the mixture takes place rapidly.
7. With the occurrence of increase in the metabolic rate during exercise, there is also an increase in the rate of gas exchange in the lungs.
8. Clarity in writing is my intention.

ANSWERS.

1. He examined the patient.
2. He performed the tests adequately.
3. These agents act by inhibiting the synthesis of cholesterol by the liver.
4. We analyzed each specimen at least twice.
5. Our preliminary report described the techniques used to infuse fluids into the cerebral ventricles of rats.
6. Ethanol evaporates rapidly from the mixture.
7. When the metabolic rate increases during exercise, the rate of gas exchange in the lungs also increases.
8. I intend to write clearly.

E. MODIFIERS

Modifiers can be words, phrases, or clauses. They fall into one of two categories: adjectives or adverbs. Adjectives qualify nouns and pronouns, the words that serve as subjects and objects of sentences. Adverbs mainly modify verbs, but they can also modify adjectives, other adverbs, and even whole sentences. Because their position can alter the intended meaning, it is important to place modifiers properly.

Adjectives. Adjectival modifiers are easy to place and usually modify the nearest noun. One or more adjectives precede the noun, while adjective phrases follow the noun.

a little red house

a child in jeans

a child in blue jeans

a child in jeans the color of the sky

Adverbs. Adverbs behave less conventionally than adjectives and are not as easy to place. Since an adverb can modify a verb, an adjective, an adverb, and even an entire sentence, the positioning of an adverb can sometimes perplex even the most practiced writer.

Most single-word adverbs end in -*ly*, such as *lightly* and *evenly*, and usually precede the words that they modify.

We usually centrifuge samples for ten minutes.

He noted a relatively large increase in blood flow.

An exception is when the adverb modifies an intransitive verb, which is a verb without a direct object. The adverb usually follows the intransitive verb that it modifies.

He walked quickly.

The machine runs efficiently.

An adverbial modifier that modifies the entire sentence can usually be placed at either the beginning or end.

As soon as my replacement arrived, I left for my sabbatical year.

I left for my sabbatical year as soon as my replacement arrived.

If the modifier makes up a large portion of the sentence, it should follow the verb for better balance.

As soon as my replacement arrived, I left.	I left as soon as my replacement arrived.

With only one hand on the wheel, he drove.	He drove with only one hand on the wheel.

Compound verbs. Compound verbs consist of one or more helping verbs (*have, be, do,* and the like) and a participle (a verb given an *-ing, -ed*, or *-en* ending). A single adverb modifying a compound verb should immediately follow the first helping verb.

He is probably writing his thesis now.

He will probably have completed his thesis by then.

If only the participle is being modified, the number of words in the adverbial determines its placement in the sentence. If the adverbial is a single word, it immediately precedes the participle; otherwise, it follows the verb.

He is probably writing his thesis now.

His thesis has been carefully written.

The procedure has been tried time after time.

Sometimes a single adverb can take one of several positions in the sentence, depending on where you wish to place emphasis. An adverb should never, however, separate a verb from its object.

Slowly, he drew the blood into the syringe

He slowly drew the blood into the syringe

He drew the blood slowly into the syringe

He drew the blood into the syringe slowly

but never

He drew slowly the blood into the syringe.

An adverbial can have more than one logical position in the sentence, each of which gives a different meaning. The sentence

He said when the experiment was finished he would leave

could be interpreted two ways, and should be rearranged to convey the correct meaning.

He said he would leave when the experiment was finished

depicts a man announcing his future departure. If rewritten as

When the experiment was finished he said he would leave

it could apply to a man who, after watching the completion of an experiment, suddenly realizes that it is time to go home.

A misplaced modifier (sometimes known as a *dangling modifier*) appears to modify the wrong element of the sentence, making the sentence confusing or illogical.

We purchased rats from a dealer weighing about 250 g.	We purchased rats weighing about 250 g from a dealer.
While at the lecture, Dr. Smith took the records from my laboratory.	While I was at the lecture, Dr. Smith took the records from my laboratory.
Oxygen content was determined immediately after collecting the samples.	Oxygen content was determined immediately after the samples were collected.

In each of the preceding examples the modifier appears to apply to the wrong word and must be moved. The first example states that the dealer weighed only 250 g. The second example implies that Dr. Smith had the ability to be in two places at once, the lecture hall and the laboratory. Finally, the third example says that oxygen collected the samples.

The following two versions of a report illustrate the importance of careful sentence construction. The first version is filled with misplaced modifiers (italicized) that make the text awkward and ambiguous. These errors are corrected in the second version, which is much easier to read. (Courtesty of M. Zeiger.)

VERSION 1:

Residual lung volumes were determined in 20 normal subjects, *using* a modified Collins spirometer with a 9-liter capacity. The following procedure was used. Before *testing* each subject, the water level of the spirometer *was checked* and if necessary restored to a predetermined level to maintain a constant dead space. After *checking* for leaks in the system, the spirometer, tubing, and breathing valve were flushed repeatedly with oxygen until the nitrogen was cleared from the system. Nitrogen clearance was monitored *using* a nitrogen analyzer. When completely *cleared* of nitrogen, 5.915 liters of oxygen were introduced into the spirometer. *Combining* this oxygen with the 1.085 liters of oxygen in the dead space, the total effective spirometer volume was 7 liters.

All tests were conducted *in the sitting position.* After *explaining* the purpose of the test and *describing* the procedure, the noseclip *was secured* firmly on the subject's nose and the mouthpiece was positioned comfortably in his mouth. Next, *turning* the breathing valve to the outside air, the subject was instructed to inhale deeply and then to exhale as fully as possible.

VERSION 2:

To determine the residual lung volumes in 20 normal subjects, *we used* modified Collins spirometer that had a 9-liter capacity. Before *testing* each subject, *we checked* the water level of the spirometer and, if necessary, restored it to a previously determined level in order to maintain a constant dead space. *We then checked* the system for leaks and flushed the spirometer, tubing, and breathing valve repeatedly with oxygen until the nitrogen was cleared from the system. Nitrogen clearance was monitored *by* a nitrogen analyzer. When the system was completely *cleared* of nitrogen, *we introduced* 5.915 liters of oxygen into the spirometer. This *oxygen combined* with the 1.085 liters of oxygen in the dead space to yield a total effective spirometer volume of 7 liters.

All tests were conducted on *seated subjects*. After *explaining* the purpose of the test and *describing* the procedure, *we secured* the noseclip firmly on the subject's nose and positioned the mouthpiece comfortably in his mouth. *We then turned* the valve to the outside air and instructed the subject first to inhale deeply and then to exhale as fully as possible.

F. GERMANIC CONSTRUCTION

Sentences containing several adjectives in sequence are difficult to understand. These sequences are sometimes called Germanic constructions, after the German practice of concatenating several adjectives and nouns to form a single, very long word. They are no easier to understand in English than they are in German! This is a case where the most concise sentence is not the clearest; a few extra words and punctuation marks will make the sentence easier to understand.

Simian virus transformed fetal mammalian heart fibroblast	Simian virus-transformed fibroblast from fetal mammalian heart
Whole rat liver homogenates were used for preparing the antigen.	The antigen was prepared from the homogenized whole liver from rats.
The gas analyzer sampling tube is then connected to the calibrator mixing chamber.	The sampling tube of the gas analyzer is then connected to the mixing chamber of the calibrator.
Respiratory rates were measured with a Yellow Spring Instrument Co. oxygen monitor system.	Respiratory rates were measured with an oxygen-monitoring system (Yellow Spring instruments Co.).

The samples contained high molecular weight complement-fixing serum antibodies.	The serum samples contained complement-fixing antibodies of high molecular weight.
The oxygen-inhalation arterial blood oxygen tensions of the neonates were consistent with this hypothesis.	The measurements of oxygen tension of arterial blood taken while the neonates breathed oxygen were consistent with this hypothesis.

G. PUNCTUATION

Remember that punctuation and spacing are important, too. Sometimes the omission of a single mark of punctuation can cause confusion. For instance, in the book title,

The Physical Chemistry of Poly(γ-benzyl-L-glutamate) and Glutamic Acid Containing Polypeptides

a reader might erroneously infer that polypeptides are sometimes found in glutamic acid. The author is actually discussing poly(g-benzyl-L-glutamate) and other polypeptides that contain glutamic acid. A hyphen is needed between *glutamic acid* and *containing* to indicate this.

The Physical Chemistry of Poly(g-benzyl-L-glutamate) and Glutamic Acid-containing Polypeptides

The following newspaper excerpt illustrates the importance of proper spacing:

In August 1993, a dam in a remote western province of China burst and killed 257 people. However, the U.N. disaster relief agency misread a Chinese document and reported the death figure at 1,257. The error resulted from a misplaced space in the English translation of the document, which read "as of September 1,257 people were dead," instead of "as of September 1, 257 people were dead."

H. AMERICAN AND BRITISH STYLES

British writing is different from American writing in certain forms of punctuation and spelling. Whatever style is used will not normally affect the reader's understanding of the text, but you should be consistent and employ the same style throughout a work. If you submit a manuscript to an English journal, its editorial office will automatically convert the punctuation and spelling to British style. Similarly, an American journal will impose the American style.

1. Spelling. Some American words ending with *-ction, -ense, -er, -ll, -log,* or *-yze* are spelled differently in British usage.

Ending	American	British
-ction	connection	connexion
	inflection	inflexion
-ense	defense	defence
	practice	practice (noun)
		practise (verb)
-er	center	centre
	liter	litre
	meter	metre (unit of measure)
		meter (instrument)
-or	behavior	behaviour
	color	colour
-l	distill	distil
-log	catalog	catalogue
-yze	analyze	analyse
	catalyze	catalyse

For some verbs ending with *-e*, the American style is to drop the silent *e* when a suffix is added, while British style retains the *e*.

Verb	American	British
acknowledge	acknowledgment	acknowlegement
age	aging	ageing
judge	judgment	judgement

The digraphs *ae* and *oe* in words of Latin or Greek derivation are retained in British style.

Digraph	American	British
ae	anesthesia	anaesthesia
	cesium	caesium
	diarrhea	diarrhoea
	etiolate	aetiolate
	hematology	haematology
	leukemia	leukaemia
oe	edema	oedema
	esophagus	oesophagus
	estrogen	oestrogen
	fetus	foetus

2. Punctuation with quoted material. The British rule for placing a comma, an ellipsis, an exclamation point, a period, or a question mark is: If it belongs with the quoted material, it goes inside the quotation marks; otherwise, it goes outside. If the quotation appears at the end of a larger sentence, the punctuation mark serves to terminate both.

Lincoln started with, "Fourscore and seven years ago," then paused before continuing.

"Fourscore and seven years ago ...."

He shouted, "Have a safe trip!" as we drove away.

She closed with, "This meeting is now adjourned."

Why did she say, "Call me when you get home"?

Colons and semicolons are placed outside the quotation marks. If the quotation ends with a colon or semicolon, it is dropped.

It was clear that everyone had read "Treatment of Tumors"; the ensuing discussion ws brisk and informed.

The American rules for placing colons, ellipses, exclamation points, question marks, or semicolons are the same as the British rules. A comma or period, however, is always placed inside the closing quotation mark. If the quote is included within another sentence, a terminal period is omitted or replaced by a comma, unless the quote comes at the end of the sentence. If the quotation appears at the end of a larger sentence, the period is placed inside the closing quotation mark and serves to terminate both quote and sentence.

"This meeting is now adjourned," were her closing words.

"The pen is mightier than the sword" is his favorite maxim.

Finally, the American practice is to use a comma after *e.g.* and *i.e.,* while the British style omits the comma.

AMERICAN	BRITISH
Please bring some form of identification, e.g., a driver's license or passport.	Please bring some form of identification, e.g. a driver's license or passport.

She had the credentials, i.e., a degree from a top-tier school and extensive experience.	She had the credentials, i.e. a degree from a top-tier school and extensive experience.

Note that scholarly Latin such as *e.g., i.e., et al.,* and so forth can be set in roman type; italics are unnecessary.

3. Block quotations Quotations that are displayed separately from the main text are called *block quotes*. Typically, a block quote starts on a new line, is indented, and may be in a smaller typesize. Because they are clearly separate from the main text, quotation marks are unnecessary.

> Every child who was educated in America is familiar with the address that Abraham Lincoln delivered at Gettysburg, Pennsylvania and began:
>
> > Fourscore and seven years ago, our fathers brought forth on this continent a new nation, conceived in liberty and dedicated to the proposition that all men are created equal.
> >
> > Now we are engaged in a great civil war, testing whether that nation, or any nation so conceived, can long endure.

A quotation that is not displayed separately must be enclosed in quotation marks. If the quotation contains more than one paragraph, quotation marks are used at the beginning of each paragraph and at the end of the last paragraph. They are not used at the ends of any paragraph except the last one.

> Every child who was educated in America is familiar with the address that Abraham Lincoln delivered at Gettysburg, Pennsylvania and began:
>
> > "Fourscore and seven years ago, our fathers brought forth on this continent a new nation, conceived in liberty and dedicated to the proposition that all men are created equal.
> >
> > "Now we are engaged in a great civil war, testing whether that nation, or any nation so conceived, can long endure."

3 Paragraph Structure

In its simplest form, a lucid paragraph contains a topic sentence and clearly related supporting sentences. The topic sentence contains the main point or idea of the paragraph, while supporting sentences provide detail or ancillary information. Each paragraph should be organized for continuity. That is, a smooth flow of ideas should be maintained, not only from sentence to sentence, but from paragraph to paragraph as well.

Less experienced authors tend to format each paragraph identically, making the first sentence a summary statement, with subsequent sentences providing the detail. Such writing "by formula" is considered less polished and often lacks good transition between topics; it is, however, direct and intelligible and therefore perfectly acceptable.

It is beyond the scope of this discussion to cover paragraph composition in detail. Authors learning to write in English would be discouraged when confronted with the many fine points of paragraph design. My advice is to write your first draft with just the goal of communicating simply and clearly; otherwise, you will hesitate to start writing at all. Then gradually improve your style in subsequent drafts. With practice you will compose more "by ear" and less by studying rules.

The following are basic guidelines for paragraph design. Readers interested in further study are referred to Chapter 3 of Mimi Zeiger's book *Essentials of Writing Biomedical Research Papers,* which discusses paragraph composition and provides many examples.

1. Cover only one main point or idea in each paragraph.
2. Each sentence should establish or support the topic of the paragraph.

Sometimes the relationship of the supporting sentences and topic is unclear, as in the following:

> Muscle length and changes in contractility have been reported to have overlapping effects on the components of excitation-contraction coupling. Muscle length is believed to affect the action potential, the amount of calcium released, and the rise of intracellular calcium...; finally, muscle length affects the interaction between actin and myosin and hence shortening and force development. Changes in contractility are believed to affect the action potential and the level and rise of intracellular calcium.

Muscle length and contractility are discussed in separate sentences and without noting their similarities. Thus the sentences do not directly illustrate their *overlapping effects,* and relationship of these sentences to the topic sentence is not immediately evident. The relationship is established by stating how effects overlap.

> Muscle length and changes in contractility have been reported to have overlapping effects on the components of excitation-contraction coupling. Both affect the action potential, the amount of calcium released, and the rise of intracellular calcium. In addition, muscle length affects the interaction between actin and myosin and hence affects muscle shortening and force development.

3. Include information that explains why actions were taken.

Although you can expect your readers to have a reasonable understanding of the topic, comprehension will be facilitated if the reader is not required to "fill in the gaps."

> All of the patient data were kept in paper files. The absence of even one clerk caused delays in the monthly reporting. Finally, management decided to interview some systems analysts.

The connection between the three sentences in the preceding paragraph is not clear. Although the meaning can be inferred, it is better to state it outright.

> All of the patient data were kept in paper files, which took much staff time to maintain. The absence of even one clerk would delay the monthly

patient reports. Management wanted computerized recordkeeping, which would take less time and be more reliable, and finally decided to interview some systems analysts to develop the new system.

4. Keep a consistent point of view.

That is, maintain the same grammatical voice (active or passive) throughout the paragraph.

Topical applications of the drug did not improve the condition. The condition improved after small doses were delivered intravenously.

The first sentence is in the passive voice, in which the condition receives the action. The second sentence is in the active voice, in which the condition performs the action. Either voice is acceptable, but the change from one voice to the other makes the logic less clear.

Topical applications of the drug did not improve the condition. Intravenous delivery of small doses improved the condition.

Or,

Unlike topical applications of the drug, intravenous delivery of small doses improved the condition.

5. Use parallel construction to make the paragraph easier to understand.

Maintain consistent structure throughout the paragraph. In an attempt to avoid monotony, some writers vary the sentence construction and thereby hinder comprehension.

A 10 mg dose produces no effect, a 20 mg dose produces a small effect, but patients demonstrate a noticeable effect from a 30 mg dose.	A 10 mg dose produces no effect, a 20 mg dose produces a small effect, but a 30 mg dose produces a noticeable effect in patients.

Part II Planning the Paper

4 Preliminaries

A. TO WRITE OR NOT TO WRITE

The application of the rules in this book cannot change the quality of a study. University professors and senior scientists in research institutes have sometimes claimed that basic research is superior to applied research, but actually there is only good research or poor research, be it basic or applied. No effort can improve the results of a flawed research plan. Your research project should start with a premise, followed by developing a theory, designing experiments, or both, to prove or disprove the premise. You should not start with data collected by your students or coworkers and then, as the armchair general, sit in your office and try to explain the data. This is simply a subjective interpretation of data obtained without plan.

There is nothing wrong, of course, with studying existing data and forming a premise from these. Just remember that this premise is the starting point, not the conclusion, of a research project. To be valid, your conclusions must be based on data resulting from experiments specifically designed to test your premise. A theoretician may develop a theory and leave the experimental proofs to others, but it is not good practice to develop theories "after the fact" to suit the experimental results obtained by others.

You must, of course, have reproducible results before even contemplating to write a paper. For statistical work, it is best to conduct a series of experiments to give reasonable assurance that your results were not due to chance. Design your experiments economically; duplicates or triplicates usually suffice.

B. FORMAT OF A REGULAR PAPER

IMRAD, perhaps the most common format for a scientific paper, stands for Introduction, Materials and Methods, Results And Discussion. A conventional manuscript consists of the following parts:

Title
Authors
Abstract
Introduction
Material and Methods
Results
(And)
Discussion
Acknowledgment
References
Tables
Figures
Legends for figures

A manuscript is rarely developed in this order. Although a finished paper can give the impression that the research project went smoothly according to design, in reality the researcher must contend with numerous obstacles and edit the text accordingly. An original idea may be modified, conflicting results must be reconciled, and some new findings may be quite unexpected.

Each author has his or her own procedure for preparing a manuscript. I normally use the following approach in preparing my own manuscripts:

1. Decide on a journal to which the manuscript will be submitted and follow the journal's Instructions to Authors
2. Decide which experimental data to present
3. Write the Materials and Methods section
4. Summarize your results, and create the figures and tables
5. Write the Introduction and Discussion
6. Write the reference list

7. Assemble the tables in numerical order, one table per page
8. List the legends for figures in numerical order
9. Assemble the figures in numerical order
10. Select a tentative title
11. Write the abstract
12. Revise the first draft immediately
13. Do not work on the revised draft for several days
14. Revise the manuscript
15. Repeat steps 13 and 14 until you are satisfied with the text
16. For a multiauthor paper, seek comments and changes from each author until all authors approve a final draft
17. Reread the manuscript, improve the sentence structure and word choice, and correct typographical errors
18. Have one or more of your colleagues review the manuscript
19. Have the text polished by someone fluent in English (If English is not your first language, this is a good opportunity to improve your writing)
20. Submit the manuscript to the journal

Scientific writing emphasizes *brevity* and *clarity*. This is especially critical when the journal imposes a limit on length. For instance, *Nature* restricts items in its "Letters to Nature" column to 1,000 or fewer words of text and four or fewer display items. *The Proceedings of the National Academy of Sciences of the United States of America* only accepts research papers that do not exceed five printed pages, including figures and tables, regardless of whether the author is a member. Thus, the saving of every word counts.

It is good practice to condense your manuscript as much as possible, even if submitting to a journal that does not restrict length. It is often possible to reduce a manuscript by 10% or more, and the final, concise manuscript is usually more lucid and readable.

There is a convention for what verb forms are used in each section of the manuscript. The Introduction tells *what* has been previously published in the literature and *why* the study was done; its verbs are usually in the present tense. The Materials and Methods section describes *how* the study was conducted; its verbs are in the past tense. The Results section summarizes *what* was found and is also in the past tense. Finally, the

Discussion explains the results and is customarily written in the present tense.

C. COMMUNICATIONS TO THE EDITOR

A *Communication* or *Letter to the Editor* is an urgent report of unusual results involving new and significant insights that allow for rapid publication. Communications and Letters are *not* designed to be short versions of regular papers; they are findings whose importance to current research warrants immediate publication. The manuscripts are required to be very brief, making conciseness and clarity even more important. The results in a Communication or Letter can be later republished in greater detail as part of a regular paper.

Some journals still accept short, complete papers and publish them as Notes. Unlike a Communication or Letter, a Note should not be republished later in a regular paper. If you have several short notes on the same subject, they are far better consolidated and published once as a regular paper.

D. NOMENCLATURE AND STYLE

International standards on nomenclature should be observed. *The Proceedings of the National Academy of Sciences of the United States of America 92 (1995)* recommends the following guides for papers published in *Proceedings* (reproduced here with permission):

General. *Scientific Style and Format: The CBE Manual for Authors, Editors, and Publishers* (1994) 6th Ed. Council of Biology Editors, South LaSalle Street, Suite 1400, Chicago, IL 60603.

Chemistry. *The ACS Style Guide: A Manual for Authors and Editors,* ed. Dodd, J. S. (1986) American Chemical Society Publications, 1155 16th Street, N.W., Washington, DC 20036.

Genetics. *Bacterial.* Demerec, M., Adelberg, E. A., Clark, A. J. Hartman P. E. (1966) *Genetics* **54**, 61–76. *Human.* Klinger, H. P., ed. (1993) *Genome Priority Reports* (Karger, Basel), Vol. 1. *Mouse.* Committee on Standardized, Genetic Nomenclature for Mice (1994) *Mouse*

Genome **92**. *Plant.* Commission on Plant Gene Nomenclature (1994) *Plant Mol. Biol. Rep.* **12**, Suppl. 2. *Other.* O'Brien, S. J., ed. (1993) *Genetic Maps: Locus Maps of Complex Genomes.* Cold Spring Harbor Lab. Press, Plainview, NY), 6th Ed.

Immunology. For human immunoglobulins and their genetic factors, the rules of the World Health Organization [or the *Council of Biology Editors Style Manual*].

Life Sciences. *Biochemical Nomenclature and Related Documents* (1992) (a compendium of IUPAC-IUB documents, all of which have appeared elsewhere) Portland Press Ltd., 59 Portland Place, London W1N 3AJ, U.K., on behalf of the International Union of Biochemistry and Molecular Biology (in North America, Portland Press, Inc., P.O. Box 2191, Chapel Hill, NC 27515-2191). This compendium contains the International Union of Biochemistry rules of nomenclature for amino acids, peptides, nucleic acids, polynucleotides, vitamins, coenzymes, quinones, folic acid and related compounds, corrinoids, lipids, enzymes, proteins, cyclitols, steroids, carbohydrates, carotenoids, peptide hormones, and human immunoglobulins.

Enzymes should be identified by the recommended name followed in parentheses by the systematic name and the Enzyme Commission (EC) number on first mention, in both the abstract and the text. For guidance refer to: *Enzyme Nomenclature: Recommendations (1992) of the Nomenclature Committee of the International Union of Biochemistry* (1992) Academic Press, New York.

Mathematics. *A Manual for Authors of Mathematical Papers* (1970) [reprinted with corrections (1980)] American Mathematical Society, 321 South Main Street, P.O. Box 6248, Providence, RI 02904.

Physics. *AIP Style Manual* (1990) American Institute of Physics, 335 East 45th Street, New York, NY 10017.

Psychology. *Publication Manual of the American Psychological Association* (1994) 4th Ed. American Psychological Association, 750 First Street, N.E., Washington, DC 20002-4242.

5 *Title and Running Title*

The past decade has seen a huge increase in the number of scientific publications. One might even call it a literary explosion. While this is a favorable indication of growth in research, it also means that today's scientist is inundated with more papers than he or she can ever hope to read. Each field has several prestigious, must-read journals. As a biochemist, I always read *Biochemistry, Journal of Biological Chemistry, Journal of Cell Biology, Journal of Molecular Biology, Nature, Proceedings of the National Academy of Sciences of the U.S.A., Protein Science,* and *Science.* There is computerized search to find papers by topic or author. There are also index periodicals such as *Current Contents,* which list only the tables of contents of journals, usually at about the same time that the journals are published. *Current Contents* is published weekly in several editions, each covering different fields of interest.

How does a scientist choose from among so many offerings? Typically, the reader will identify interesting papers by scanning the list of titles and authors from a journal's table of contents, an index periodical, or a computer search. Most titles will receive only a fraction of a second's consideration this way. Papers that do not pass this initial screening may not be read.

Clearly, it is vital to choose a title that will interest the reader by providing specific information in as few words as possible. In that sense it is a commercial or an advertisement that will interest the audience in reading the paper.

A title is usually a phrase, but can be a complete sentence. In composing a title for a paper you should:

1. Provide specific information in as few words as possible.

2. Be informative and lucid.

3. Include a subtitle, if further detail is needed.

Induced conformation of some peptide hormones in lipid solutions	Lipid-induced conformation of some peptide hormones: preponderance of α-helices

The subtitle in this case narrows down the type of conformation.

4. If the work concerns changes, the positive or negative direction of the effect or action should be specified.

Effect of amino acids on the sickling of erythrocytes containing hemoglobin S	Reversal effect of amino acids on the sickling of erythrocytes containing hemoglobin S

The preceding right-hand title came from an actual paper that was later proven to be incorrect. The researcher who challenged the findings published under the title,

Do amino acids reverse the sickling of erythrocytes containing hemoglobin S?

Note that the question format subtly implies a doubt of such effect, thereby indicating the findings presented in the paper.

5. Avoid nonstandard abbreviations.

Some standard abbreviations such as DNA, RNA, and ATP are so widely used that they are rarely spelled out in full. Bear in mind, however, that an abbreviation common in one field may not be recognized by all readers of a multidisciplinary journal.

The CMC of SDS	The critical micelle concentration of sodium dodecyl sulfate

IUPAC discourages the use of the same letter to represent two different things, such as S for both *sodium* and *sulfate,* or C for both *critical* and *concentration.* In conversation, however, researchers usually opt for convenience; SDS is undeniably easier to pronounce than NaDodSO4. The abbreviation SDS has gained acceptability through popular usage, and some journals now allow the use of either SDS or NaDodSO4. Remember, though, that it is never incorrect to follow IUPAC standards.

6. Begin with an important term to give immediate impact.

Bear in mind that the reader tends to focus on the first few words of the title, so terms that convey key information should be at the beginning. The corollary is to avoid beginning the title with a general word such as *The, A, Results,* or *Study.* Any scientific paper is understood to be a study or an investigation, which produces results. The words *study* and *results* do not add information and should therefore be avoided in the title.

The investigation of the relationship between sudden deafness and atherosclerosis	Sudden deafness and its relationship to atherosclerosis
Studies on the binding site of *Vicia villosa* lectin	*Vicia villosa lectin* and its carbohydrate-binding site

7. Avoid subjective evaluations.

Subjective terms such as *novel* and *innovative* are better omitted; the reader is the appropriate one to make such judgments.

A novel method for determining the molecular weights of denatured proteins	A rapid method for determining the molecular weights of proteins: intrinsic viscosities in 6 M guanidine hydrochloride

8. Avoid serial titles.

Many journals no longer allow the use of serial titles. For example,

Polypeptide XIV. Helix-coil transition of poly(L-ornithine)

would be edited to

Helix-coil transition of poly(L-ornithine)

If your paper is part of a series, this can be indicated by a footnote to the title.

9. Check the Instructions to Authors for any other rules.

Nature, for example, specifies that active verbs, numerical values, abbreviations, and punctuation should be avoided.

10. Provide a running title.

Most journals now require a running title, usually less than fifty characters long including spacing. This is the briefer title that appears on each page of text, usually at the top; thus, it is also known as a *running head.* (Perhaps a running title printed at the page bottoms can be humorously called *running feet.*) Using some of the preceding examples:

Lipid-induced conformation of some peptide hormones: preponderance of α-helices
Running title: CONFORMATION OF PEPTIDE-LIPID COMPLEXES

The critical micelle concentration of sodium dodecyl sulfate
Running title: CRITICAL MICELLE CONCENTRATION OF NaDodSO$_4$

Sudden deafness and its relationship to atherosclerosis
Running title: SUDDEN DEAFNESS AND ATHEROSCLEROSIS

Vicia villosa lectin and its carbohydrate-binding site
Running title: LECTIN-CARBOHYDRATE COMPLEXES

A rapid method for determining the molecular weights of proteins: intrinsic viscosities in 6 M guanidine hydrochloride
Running title: VISCOSITY AND MOLECULAR WEIGHT OF PROTEINS

6 Authors

The byline appears on the title page of the manuscript. It follows the title and presents the full names and institutional affiliations of all authors. Any additional information (such as biographical data), including footnotes, should be placed on a separate page.

A. MULTIAUTHORSHIP

Naming the author is a simple matter when a researcher works independently, but this is rarely the case these days. The task can become complicated when the research project involves more than one worker, such as a research supervisor or other significant contributors. Who should be listed as an author? In what order should coauthors be listed? In some cases, a coauthor may have only contributed peripherally to a work and may not even have read the final manuscript. Occasionally, researchers approach collaborative work as a sort of mutual admiration club: "I will list your name in my paper, and you will list my name in your publications." This practice should be discouraged.

The reader of a multiauthor paper tends to notice the first author, who usually did the research, and the last author, who was probably the supervisor or the laboratory chief. I have occasionally seen publications based on Ph.D. theses, in which the supervisor listed himself as senior and the graduate student or postdoctoral fellow who actually did the work as coauthor. One may question in such cases whether the graduate student had fulfilled the requirements for an advanced degree, since his contribution did not earn him the position of first author. I also knew of a distinguished scientist who often listed himself as the last author, yet occasionally in an oral presentation he could not answer questions from the floor about some experiments done by his coauthors. There are occasional attempts to exploit the author list and garner attention by naming a distinguished scientist as the last author (sometimes even as the first author), regardless of whether the scientist fully understands the contents of the paper.

Some authors wish to reward unusually gifted or devoted technicians by listing them in the byline. This is accomplished by adding, after the names of authors, "With the technical assistance of ..." Some journals discourage this practice, but others permit it, thereby making the situation bibliographically messy. I believe that the title page should contain only the names of those who have contributed materially to the work. Those who have participated in an advisory or supporting capacity can be thanked in the Acknowledgments.

Today there is a proliferation of publications with a disproportionately large number of coauthors. To cite such a paper can be quite cumbersome. Some journals now reduce a list of multiple authors, usually more than three, to the first author's name followed by "et al."

I was quite taken aback, and perhaps you will be, too, by the following news items. The first appeared in the American Chemical Society's *Chemical & Engineering News* (September 28, 1981).

A scientific paper with 77 authors [in *Nuclear Instruments & Methods*] was mentioned here recently (*C&EN*, July 27, page 108), with the comment that it must set some kind of record. Not so, says A. Eastwood of Chalk River, Ont., Canada. He cites a paper with 89 authors that appeared in Nuclear Physics [**B52**, 414 (1973)] Also, he notes, the number of authors of either paper exceeds the number of words used by Philip Abelson to confirm the discovery of nuclear fission [*Phys. Rev.*, **55**, 418 (1939)]. The announcement said:

We have been studying what seemed to be L x-rays from the seventy-two-hour 'transuranic' element. These have now been shown by critical absorption measurements to be iodine K x-rays. The seventy-two-hour period is definitely due to tellurium as shown by chemical test, and its daughter substance of two-and-a-half-hour half-life is separated quantitatively as iodine. This seems to be an unambiguous and independent proof of Hahn's hypothesis of the cleavage of the uranium nucleus.

"You have to count the words in a particular, but legitimate, way to arrive at fewer than 77," according to Eastwood. "Anyway," he also says, "It is not surprising that Dr. Abelson went on to become editor of *Science*." In *Chem. Abstr.* **33**, 2800[8] (1939), Dr. Abelson's abstract reads:

> What seemed to be *L* x-rays from the 72-hr, trans-U element were found to be I *K* rays. The 72-hr. period is due to Te (chem. test); the product with 2.5-hr. half life is I (chem. sepn.). This is proof of Hahn's hypothesis (*C.A.* **31**, 6552[7]) of the cleavage of the U nucleus.

The second news item appeared in *Random Samples* of *Science* in its 16 September issue (volume **241**, page 1437) and 15 November issues (volume **242**, page 1130) of 1988.

> A 1988 paper in *Phys. Rev. Lett.* entitled "Experimental mass limit for a fourth-generation sequential lepton from $e+e^-$ annihilations at 1 s = 56 GeV" (volume **61**, pages 911–914) listed 104 authors from 19 universities, most of them in Japan. A companion piece seems a bit skimpy with only 75 authors from 18 universities. The list of authors was so long it forced the two-page table of contents into a third page. The *Chem. Abstr.* only listed nine authors followed by *et al.* One then wonders whether this is the record number of authors on a publication. The National Library of Medicine, Washington, DC turned up a 1986 paper in *Kansenshogaku Zasshi* entitled "Comparative study of MI-0787/MK-0791 and piperacillin in respiratory tract infections," which had 193 Japanese authors from 20 institutions. Another 1986 paper in *Plasma Phys. Controlled Fusion* entitled "Confinement and heating of plasmas in the JET Tokamak" sported 246 authors. The JET group had another paper by 257 authors, which, however, was not in a peer-reviewed journal.
>
> In a 1987 paper in *Phys. Rev. D.* there were 108 authors from 14 universities, most of whom were Americans. Therefore, Americans seem to be in the running against Japanese, and multiple authorship is one area in which no country has yet established a monopoly. *Science* has raised a question of whether we need an international Author Nonproliferation Act. Normally all the authors of a quoted paper should be listed in the references of a manuscript, whereas they can be abbreviated as the family name of the first author followed by *et al.* in the text. I would, however,

make exceptions in these cases to listing all the authors in the References just to save the space of a journal. Have we reached a record in the above examples and can anyone top 246 (peer-reviewed) or 257 authors?

B. FORMAT OF CHINESE NAMES

In Western countries a personal name is written in the following order: first name, followed by middle names or initials, then last name. It is assumed that the last name, which may be hyphenated, is the family name. Occasionally the family name is followed by designations such as *Jr.* (for *junior,* a man with the same name as his father), *Sr.* (for *senior,* a man with the same name as his son), M.D., F.R.S. (Fellow of the Royal Society), and so on. These are easily recognized as not being part of the family name.

In China, Japan, and many other countries, the traditional style is to list the family name first, followed by the given names. If the name is Westernized, as it often is by authors writing in English, the order is reversed.

TRADITIONAL	WESTERNIZED
Qi Zixian	Zixian Qi
Samejima Tatsuya	Tatsuya Samejima
Shen Lu-ping	Lu-ping Shen
Yang Li	Li Yang

Most Chinese have a one-character family name and a one- or two-character given name. Those familiar with Chinese names will immediately recognize that in the name *Shen Lu-ping,* Shen is the family name and Lu-ping is the given name. The family name for *Yang Li* is ambiguous, however, because both names are single-character.

Asian names present a real challenge for the indexer, who may not know which word represents the family name. To complicate the matter further, I have even seen one Chinese author named in both traditional order and Western order in the same English-language paper. Perhaps out of cultural pride, Chinese authors in English journals published in China (except for the province of Taiwan) insist on listing their names in the traditional way. This is fine for well-known names, but it will be a nightmare for the indexer if you are ordinary people.

For a paper written in English, the Western style should be used for all personal names. It is pointless to undermine the effectiveness of your paper by interjecting cultural politics. As the proverb says, "When in Rome, do as the Romans do." Chinese sciences are in the process of catching up with Western, and by the twenty-first century Chinese may become an international language. When papers are published in Chinese, then the editors can set the standards on whether the author John Smith should be named as *Smith John*.

C. ROMANIZATION OF ASIAN NAMES

Names must be *romanized* or transliterated into Latin letters for publication in an English-language journal. Romanization can be tricky because some foreign sounds cannot be exactly reproduced in English. Also, as you have doubtless learned, English rules of pronunciation are by no means consistent. As a result, a name may be spelled several different ways in English. For consistency, employ a widely-accepted romanization system.

1. Chinese names.

In 1956 China introduced the romanization system called *pinyin* to replace the older Wade-Giles romanization system and place-name spellings in Webster's *Postal Atlas of China*. Pinyin phonetics are based on the official national dialect, whereas many of the Wade-Giles and *Postal Atlas* words reflect outmoded pronunciation (for example, *Beijing* versus *Pei-p'ing* and *Peking*). Pinyin is also more systematic, eliminating the ambiguities and confusing punctuation inherent in Wade-Giles.

WADE-GILES/*POSTAL ATLAS*	PINYIN
Chou	Zhou
Peking	Beijing
Tsing	Qing

Some older romanized names are not converted to the pinyin system. Premier Chou En-Lai is not usually written as Zhou Enlai, and Peking

University and Tsinghua University retain their old names.

The spelling of a name in English is a matter of personal preference. I see no problem in spelling each person's name however he or she prefers (unless cultural politics limit your choice). My given names, for example, are spelled *Jen Tsi*, not *Jen-Tsi, Jen-tsi,* or *Jentsi.* I have kept this spelling rather than switching to the pinyin spelling *Renji.* I also include a space between my given names so that my abbreviated name can be indexed as *Yang, J. T.* This helps differentiate me from what might be many listings of Yang, J.

When referring to a person who has more than one name, the general rule is to use the most recognizable name. For example, you would refer to François-Marie Arouet by his pen name Voltaire. Indexing problems arise when both names are equally well-known, such as scientists in China who formerly published under their Wade-Giles names but are now known by their pinyin names. In such cases you should use the author's current professional name, which for clarity may be followed by other names in parentheses. When citing a publication in your text, you must of course spell the author's name as it appeared in the work. The bibliography usually contains both names, for the convenience of the reader.

> The first promising results (Qi, 1946) were followed by a decade of experiments by Yang (née Li), who presented two major papers in 1953 and 1957.
>
> Li, Yee-Mui [later Yang, Yee-Mui] (1953) ...
>
> Qi, Zixian [Ch'i, Tzu-hsien] (1946) ...
>
> Yang, Yee-Mui [Li, Yee-Mui] (1957) ...

Née is optional and indicates the maiden name of a woman who has taken her husband's name.

2. Japanese names.

Japanese names are usually romanized according to the system presented in *Kenkyusha's New English-Japanese Dictionary,* 5th ed. 1980.

3. Other names.

To transliterate names from other countries, see *ALA-LC Romanization Tables: Transliteration Schemes for Non-Roman Scripts.* Randall Berry, Ed. 1991. This publication is available from the Library of Congress Cataloging Distribution Service, Washington, DC.

7 Abstract and Key Words

Most regular articles begin with an informative abstract. Unlike an indicative summary, which describes what *will be* covered in the paper (much like a table of contents), the abstract gives actual data. It is a mini-paper that is understood on its own without reference to the paper proper. The abstract should provide maximum information with minimum words, covering the

Objective

Materials and Methods

Results

Conclusions

In other words, an abstract should answer the questions *why, how* and *what*. *Why* did you study it? *How* did you study it? *What* did you find and *what* does it mean?

Why can be omitted if the objective is clear in the title. *How* should be elaborated on only if it is a paper on methodology; otherwise, it should be very brief or even omitted if well-known. *What* should selectively include only the important findings and conclusions. Most journals limit the abstract to 150 to 250 words or even less. Thus, there is no room to waste words. As a rough guideline, your abstract should be confined to within one double-spaced typed page on standard-size paper. Standard paper in the United States is slightly wider and shorter (8½ in x 11 in) than paper used in Europe and Asia (210 mm x 297 mm).

Avoid using abbreviations in the abstract unless a term, especially a long one, is used several times. In such cases it is usual to spell out the first occurrence of the term, followed by its abbreviation in parentheses.

The University of California at San Francisco (UCSF)

Where a name is better known by its abbreviation than its full spelling, the abbreviation becomes the standard and is preferred; DNA, RNA, and ATP are notable examples. The full spelling may be added in parentheses, especially if it is used elsewhere in the text.

DNA (deoxyribonucleic acid)

Avoid citing references in the abstract, which can distract the reader because every citation in the abstract must provide complete bibliographic information, including the authors, publication date, journal, and pages.

One common mistake is to end the abstract with a reference to the main text, such as, "The results will be discussed," which really tells nothing and has no place in an abstract.

Key words. Many journals require a list of three to five key words or short phrases for indexing. Some journals further specify that words already in the title not be included.

Title: Conformation of β-Endorphin in Sodium Dodecyl Sulfate Solution

Keywords: opioid peptide, polypeptide hormone, surfactant, circular dichroism,

but not

conformation, β-endorphin, and so forth.

The following abstract is included as a note of historical interest. It dates from an unusual period for scientific publications in the People's Republic of China and is typical of the type of paper produced under the political influences. During the Great Proletarian Cultural Revolution (1966–1976), most journals initially ceased publishing and later resumed in a different format that discouraged personal glories and published manuscripts under collective authorship. The leading article was usually an editorial about the Marxism, Leninism, and Mao Zedong Thought. Scientific researchers all proclaimed to be guided by Chairman Mao's revolutionary line and inspired by the spirit of "self reliance and hard struggle." The abstract is from a paper on the synthesis of glucagon.

Protein Synthesis Group, Shanghai Institute of Biochemistry, Academia Sinica. 1975. Total synthesis of crystalline glucagon by the method of solid state phase condensation of fragments. *Acta Biochem. Biophys. Sinica* 7:119–138.

Abstract: Chairman Mao taught us, "One should seriously sum up one's experience." Looking back at the experiences of our own and of others in the past decade or so on the total synthesis of proteins and polypeptides, we have analyzed the inherent contradictions of the two alternative routes of synthesis on the basis of the dialectical viewpoint of "one divides into two." Either the solid phase or the solution synthesis is fraught with difficulties when the target exceeds 100 amino acids. A new synthetic strategy was developed which we believe could resolve the contradictions inherent in the synthesis of large peptides—the solid phase stepwise condensation of peptide fragments instead of amino acids. Following repeated practice guided by the spirit of independence and self-reliance, we succeeded in synthesizing first of all a new supporting medium which provisionally met our requirements. We shall describe in the present paper the successful total synthesis of the nonacosapeptide glucagon using this supporting medium.

The paper was published in Chinese with an English abstract, of which the preceding paragraph made up about half. Today, Chinese research philosophy is the same as that of Western countries. Under market economy scientists are also encouraged to market the fruits of their research, providing partial support for their studies.

8 Introduction

An introduction contains material that should be read before the rest of the paper. Its purpose is to provide background information that the reader needs to understand the research project. Someone who feels inadequately prepared for a paper is unlikely to read it, so it is to your advantage to acquaint the reader with the subject. An introduction typically includes the reason for undertaking the project, relevant findings, and specialized background facts.

The reader is assumed to have a basic familiarity with subject. Thus the introduction excludes elementary facts and presents information relevant to the paper that only a specialist would be expected to know. The introduction must clearly specify the nature and scope of the problem studied or the questions addressed. It includes a brief summary of previous work in the field to bring the reader up to date on the topic. Naturally, you can summarize your own previous work, but an introduction is not a place to showcase your talents.

Today reviews such as *Annual Review of Biochemistry* cover much of the information that used to be provided in the introduction. Extensive background information is no longer necessary in the introduction, because researchers are expected to keep current in their fields.

The introduction aims to evoke interest and should also be brief to avoid losing the reader's attention. If the writing is clear and concise, two or three paragraphs will usually suffice. Essentially, the introduction covers three parts: *(a)* the general background; *(b)* previous findings by others; and *(c)* your examination of the questions addressed. You can very briefly describe your experimental or theoretical approach, and perhaps the principal findings. Since these are presented in the main body of the paper, they can be omitted here unless your choice of methodology requires explanation.

Verbs in the introduction are usually in the present tense for ongoing truths and others' findings, but in the past tense for your own methods and findings in the research project.

> Doctors recommend taking aspirin each day to prevent heart attacks and strokes. We administered daily aspirin to 200 subjects and found it unsuitable for ulcer-prone patients.

> We used amoxicillin, even though it has a short shelf life, and verified the pediatricians' claim that it causes fewer side effects.

The introduction often includes a literature search (list of publications on the topic) to provide the reader with alternative reference materials and suggestions for further study. In a biochemistry paper, for example, a list of sources might include

Books and articles

Index periodicals such as *Current Contents* and *Chemical Titles*

Abstract journals such as *Chemical Abstracts* and *Biological Abstracts*

Colleagues and others

Computer search

Computer Search. As library card catalogues are replaced by computer databases, it becomes much easier to compile a source list. Computers can rapidly search a library's entire holdings by criteria such as author, title, subject, and so forth. Libraries belonging to OCLC (On-line Computer Library Center) can, through the Internet, search the holdings of other member libraries as well.

Many libraries subscribe to CD-ROM (compact disc read-only memory) database services that provide information on articles in biomedical and high-impact clinical journals, such as the *New England Journal of Medicine* and the *Journal of the American Medical Association.* The computer allows you to specify search criteria and provides a list of papers meeting these criteria. The results of a computer search can be printed to paper or downloaded to diskette. Bibliographic databases contain only the citations: Title, author, date, subject, and key words. More extensive databases include the abstract or even the entire text of the paper. Because most CD-ROM databases are updated quarterly, journal articles lag behind 45–90 days from their time of publication.

There are many CD-ROM databases available from competing vendors. They differ primarily in the publications that are included, the amount of information carried for each article, frequency of update, and search features. The American Chemical Society, for example, supplies other vendors with the bibliographical information in *Chemical Abstracts* but retains the abstracts for their own database products.

Note that a CD-ROM database contains only a fraction of the journals in publication. Mainframe databases can store more extensive electronic libraries, including entire journals. The more sophisticated systems allow you to browse through a journal page by page, skim tables of contents, and see the graphics and photographs as they appear in the print journal. At present such systems are only available at large research institutions or through the Internet.

Abbreviations and Footnotes. To save space, abbreviations are often used throughout the paper and in footnotes. The rule of thumb is that a term is abbreviated if it will be used five or more times, and abbreviations should be defined at the first mention of the term, then used throughout the remainder of the text.

Footnotes in manuscripts are often listed on a separate page. They should be kept to a minimum because they interrupt the flow of reading and can create layout problems when the paper also has exhibits. If possible, include the footnote information in the text between parentheses.

9 Materials and Methods (Experimental Procedures)

The Materials and Methods section describes your experimental procedures. Unless your entire paper is on methodology, it is perhaps the easiest part of a manuscript to write, because you do not interpret data or reach conclusions. It is a straightforward recounting of your approach. This section should not be too lengthy; some information may be of lesser interest and can be de-emphasized by using smaller print (as practiced by the *Journal of Biological Chemistry*) or by placing it in a supplemental-material section (*Biochemistry*). Sufficient information should be provided so that an interested reader can repeat the experiments. When deciding what information to include in Materials and Methods, bear in mind that a careful reviewer will read this section to judge whether your procedures were sufficient enough to provide a valid answer to the question raised in the Introduction.

The name, source (lot number, if any, for commercial chemicals), purity, and potency of each material used should be stated. *Whenever hazardous materials and dangerous methods were used, the necessary precautions should be stated.* Experimental conditions are briefly but precisely described. Novel methods should be described in detail, but published methods should merely be cited by appropriate references to both the original and any published modifications. Any major modification of a well-established procedure must be detailed. The methods of determining quantitative measurements (for example, the concentration of a protein in solution) should be mentioned so that the reader can be aware of possible inherent errors. This information is often overlooked in the publications, making it difficult to guess the accuracy of the quantitative measurements.

Biochemistry requires its authors to make available to academic researchers for their own use any materials reported in the paper that are not otherwise obtainable. Such requests should, however, avoid conflicts of competition with the original laboratory.

Verbs in this section are generally written in the past tense. Use of the passive voice is also acceptable, especially as a means of placing emphasis. For instance,

Enzyme A was purchased from Sigma

is better than

We purchased enzyme A from Sigma

because emphasis here should be on the material, not the authors.

10 Results

Results are general statements that interpret the raw data obtained from experimental measurements. The Results section is the meat of a paper, the most important part of a study. All other sections serve subordinate roles, either preparing the reader for the Results, or providing supplemental information to augment the findings.

The results are presented as text, illustrations, and tables. All three forms may be used, but the same data should not be repeated in more than one form. This chapter discusses the text only; illustrations and tables are covered in chapters 11 and 12.

The text may be any length. Sometimes a statement as brief as, "The results are shown in Figures 1–4 and Tables I–III," is sufficient. For clarity, long passages of text are often organized by topic into subsections, with a subheading for each topic. The subheadings help the reader locate paragraphs that are of personal interest.

Sometimes the Results and Discussion are combined into one section. This is particularly useful when preliminary data must be discussed to show why subsequent data were taken.

Whatever format you adopt, the following guidelines should be observed:

1. Emphasize only important observations that will answer the question or solve the problem raised in your Introduction. Remember that this is not the place to publicize your research skills.

2. Be selective about your results. Too much detail can burden and even confuse the reader, who might then lose interest. Detailed data belong in a supplemental-material section (usually in smaller print at the end of a paper, to be read with a magnifying glass) or in some data bank external to the paper.

3. Structure the text so that the emphasis is on the results. Ancillary information concerning materials and methods, legends for figures, titles for tables, etc., should receive less emphasis. This can be accomplished through any of a number of writing style techniques, such as placing the results sentence at the beginning of the paragraph, putting ancillary information in a subordinate clause, or using the passive voice.

> We broke the disulfide bonds of ribonuclease A by dithioerythreitol. This reduction denatured the proteins.

The preceding example describes first the experimental procedure, followed by the result (the denaturation of the proteins) in a separate sentence. This order places more emphasis on the procedure than the result. Further, the use of the active voice accentuates the subject we over *disulfide bonds*. Although the active voice is often preferred in writing, in this case it emphasizes the wrong element. In the revision,

> When the disulfide bonds of ribonuclease A were broken with dithioerythreitol, the reduced protein was denatured

the method is described in a dependent clause, while the result receives greater emphasis as the main body of the sentence. The dependent clause is also written in the passive voice to accentuate *disulfide bonds*. Some other acceptable forms are:

> The breaking of the disulfide bonds of ribonuclease A by dithioerythreitol denatured the protein.

> Ribonuclease A was denatured by breaking its disulfide bonds with dithioerythreitol. (This version presumes the reader knows that dithioerythreitol is a reducing agent for disulfide bonds.)

One effective way to emphasize results is to state the summary results in the first sentence, with subsequent sentences providing supporting details.

The conformation of *Phaseolus caralla* lectin was stable between *p*H 4 and 10. The protein was denatured at *p*H 12, but its conformation could be restored upon lowering the alkaline pH to neutral. However, denaturation at *p*H 13 was irreversible.

4. Do not include information that properly belongs in other sections of the paper such as Materials and Methods.

5. Do not repeat the legends for figures or the titles of tables in the text.

6. Explain in the text only those illustrations and tables whose significance is not obvious to the reader. Important features that are readily apparent from the illustrations and tables should be pointed out in the text. Do not, however, repeat the data presented in the illustrations and tables.

Figure 1 shows the conformation of lysozyme as a function of temperatures. The circular dichroic spectra indicated that the protein was completely unfolded above 70°C.

Based on circular dichroic spectra, the conformation of lysozyme was completely unordered above 70°C (Fig. 1).

Or,

Circular dichroic spectra (Fig. 1) indicated that lysozyme was thermally denatured above 70°C.

Table 1 lists the percent binding of oxygen to hemoglobin as a function of the oxygen pressure. It increased with increasing pressure of oxygen.

The binding of oxygen to hemoglobin increased with increasing pressure of oxygen (Table 1).

7. Be sure that the text, illustrations, and tables are consistent with one another. This point seems obvious, but it is not uncommon to find tables with numerical values that do not agree with the figures, or data in tables or figures that are misquoted in the text.

8. Analyze your data by statistical methods, if appropriate.

9. Be honest. Do not omit data that do not support your hypothesis and conclusion or do not answer the research question.

10. Use the past tense of verbs in the Results section, except when referring to figures and tables. Use the present tense when referring to figures and tables.

>Tacrine inhibited the activity of acetylcholinesterase noncompetitively.

>Table 1 contains data collected over a four-month period.

But,

>The data in Table 1 were collected over a four-month period.

11. Terms beginning a sentence are nearly always spelled out. A sentence should not begin with an abbreviated term, unless the term is customarily abbreviated.

Fig. 1 shows ...	Figure 1 shows ...
Deoxyribonucleic acid tests are used in forensic laboratories.	DNA tests are used in forensic laboratories.
Doctor Kevin Anderson performed the tests.	Dr. Kevin Anderson performed the tests.

In the second example, the abbreviation is preferred because DNA is used far more often than deoxyribonucleic acid.

12. A sentence should not begin with a numeral or symbol. A numeral or symbol beginning a sentence should be spelled out, or the sentence rewritten.

12 subjects were tested.	Twelve subjects were tested.
1 g of compound A was used.	One gram of compound A was used. (Note that *gram* is also spelled out.)
α-tocopherol was applied.	Alpha-tocopherol was applied.
1990 was the first year the conference was held.	The conference was first held in 1990.

13. Numbers in a sentence should not begin with a decimal point. Decimal fractions less than 1 should be written with the numeral 0 before the decimal point.

.476	0.476
-.05	-0.05

11 Figures

One picture is worth a thousand words!

The text in the Results section is often supplemented by figures or illustrations, such as graphs and diagrams. (Tables, which are textual, are not considered illustrations. They are discussed in the next chapter.) The interested reader will usually, after scanning the Title and Abstract, glance over the illustrations and tables for an overview of the work. Therefore, each illustration with its legend should be self-explanatory and designed for readability.

A. DESIGN PRINCIPLES

1. Read the Instructions to Authors of the journal to which your manuscript will be submitted. Follow the journal's specified format, which determines some features of figure design. Many journals allow only lines, symbols, and numerals to identify various curves, and permit headings and descriptive explanations in the legend only. A few require that all figures be boxed. Also bear in mind that color illustrations are very costly and are discouraged by the journal unless color is essential for readability.

2. Design your illustration to fit the columns of the journal.

Most journals print the text in two columns per page. A few journals are single-column and still others, such as *Science* and *Scientific American*, are three-column. An illustration must fit within the printed area of a page and ideally will cover a whole column or columns in the horizontal direction. Figures that span only part of a column create unused white space.

Figure 1 shows an illustration that is almost as wide as the printed area of the page. The journal may decide to expand the illustration, but it could also be reduced to fit a column width, freeing up space.

Fig. 1. Folding of a protein molecule predicted by the amino acid sequence method. Each sharp peak represents a β-sheet residue and each loop an α-helix residue; the horizontal lines represent unordered residues. Computer plot by M. H. Briscoe. Reprinted, with permission, from Wu, Hasegawa, Smith, Loh, Lee, and Yang (1990) *J. Protein Chem.* **9**, 3–7. © 1990 by Plenum Publishing Corp., New York, NY.

3. Size the illustration to optimal dimensions.

Figure 2 shows a page from a book. Its illustration could be scaled to fit one column of a two- or three-column journal, but in this single-column book it leaves too much white space on the sides. It should take up more horizontal space.

Be aware, however, that when a journal reduces or enlarges an illustration, all dimensions will change proportionately. Enlarging the illustration in Figure 2 to take up more horizontal space would lengthen the vertical dimension too much. This illustration would make better use of the space if the author changed its proportions by doubling the size of the abscissa.

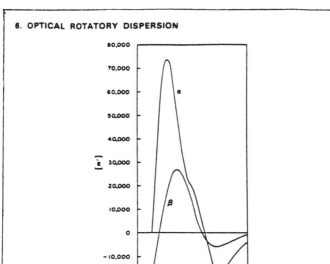

FIG. 6. Ultraviolet rotatory dispersion of poly-L-lysine at pH 11.06 ($C = 0.0123$ "$_n$). Curve α. unheated helical conformation: curve β. α heated at 51°C for about 15 min. until no further decrease in $|m|_{199}$. and then cooled to 22.5°C.

Could easily be overshadowed by those of the helices if they coexist in a protein molecule (Sec. VIII). [Circular dichroism of the β form of both poly-L-lysine and silk fibroin shows a negative dichroic band at 218 mμ; silk fibroin reveals in addition a larger positive 197-mμ band. Iizuka and Yang (26) further predict a negative band below 180 mμ as the two observed dichroic bands would give dextrorotations in the visible region, even after taking the background levorotations into consideration.] Sarkar and Doty (27) also reported that the β form of poly-L-lysine can be produced at neutral pH with the addition of sodium dodecyl sulfate, but the magnitude of the 230-mμ trough was greatly reduced compared with that in water alone (pH 11). Davidson et al. (27a) further reported that the β form of a poly-L-serine film cast from trifluoroacetic acid showed a trough at about 233 mμ and a peak at 210 mμ with a cross over near 222 mμ. The ORD profile in this case resembled that of the oriented film of the β form of poly-L-isoleucine (29).

Unlike the helical and coiled conformations, the various β forms (intermolecular parallel and antiparallel β forms and intramolecular cross-β form) may all have their characteristic Cotton effects. So far, all available

Fig. 2. Optical rotatory dispersion of poly(L-lysine) in its α-helix and β-sheet conformation. Reprinted, with permission, from Yang in *Conformation of Biopolymers,* edited by G. N. Ramachandran, 157–172. © 1967 by Academic Press, Inc., London.

4. Indicate the units of the X- and Y-axes.

For example, in Figure 2 the illustration fails to indicate the units of the Y-axis. It should specify that reduced mean residue rotation [m'] is given in deg cm^2 $dmol^{-1}$.

5. Scale units by a power of 10 so that tick marks on the axes are labeled with one- or two-digit numbers.

In Figure 2 the tick marks on the vertical axis are labeled -20,000, -10,000, ..., 70,000, 80,000. The axis should be labeled [m'] $\times$ 10^{-4} deg cm^2 $dmol^{-1}$ (alternatively, [m'] 10^{-4} $\times$ deg cm^2 $dmol^{-1}$) to allow the tick marks to be labeled -2, -1, ..., 7, 8, which are easier to read and less visually distracting.

6. Avoid wasted space within an illustration.

For instance, Figure 3 can be improved by removing the empty space below pH 1 and the straight-line data above *p*H 5 (which would be replaced by a sentence in the text).

7. Use labels, symbols, and scales large enough to be read after the illustration is reduced.

The text and symbols in Figure 3 would be difficult to read if the illustration were reduced. By doubling their size, the illustration can easily be reduced by 50% and still be legible.

8. Use standard symbols to indicate data points.

The most common ones are o , •, Δ, ▲, ▢ and ■. The symbols $\times$ and + do not stand out well and should be avoided. A curve should not be drawn through an open symbol and should be broken around a solid symbol. With smooth curves, symbols for individual data points can often be omitted.

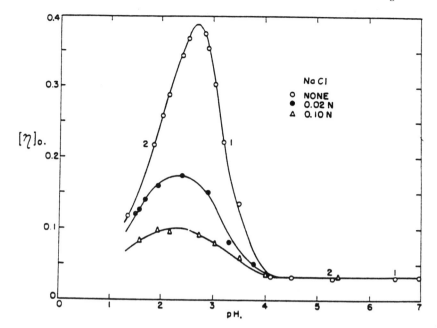

Fig. 3. Intrinsic viscosity of bovine serum albumin in acidic *p*Hs. Reprinted, with permission, from Yang and Foster (1954) *J. Am. Chem. Soc.* **76**, 1588–1595. © 1954 by American Chemical Society, Washington, DC.

9. Draw scales on all four sides of a boxed figure.

For example, the top and right axes in Figure 4 are meaningless without labeled tick marks. The values of plot points can be estimated more easily with scales all around the figure.

10. Solid plot lines are preferable, except where overlapping or crossing curves would be difficult to understand.

A dashed line is often used to represent a control or standard curve, which contrasts with solid lines for the experimental curves.

Figure 4 is an example where solid lines should not be used for all of the curves. The dashed lines make it easier to distinguish the curves near the overlaps.

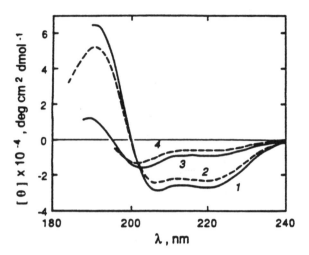

Fig. 4. Circular dichroic spectra of succinyl (solid line) and acetyl (dashed line) C-peptide analogues of ribonuclease A. Curves: 1 & 2, in 10 nM NaDodSO$_4$; *3 & 4,* in water. Reprinted, with permission, from Wu and Yang (1990) *Biopolymers* **30**, 381–388. © 1990 by John Wiley and Sons, New York, NY.

11. Do not clutter the axes with tick marks and numbers.

Excess detail is visually distracting and draws the reader's attention from the curve, as in Figure 5A. Figure 5B illustrates an improved version with fewer ticks and labels. Likewise, in Figure 2 the ordinate would be improved by changing $[m']$ to $[m'] \times 10^{-4}$ and using tick marks -2, 0, 2, 4, 6, and 8.

12. Do not crowd a graph with overlapping curves and symbols.

In Figure 6 *top* the individual curves are difficult to distinguish around 220 nm. One solution, as shown in Figure 6 *bottom,* is to remove the data point symbols. This is permissible because with repeated trials the data approach a smooth curve and symbols for individual data points become unnecessary. Further, the number of curves can be reduced in the overlapping section; two or three similar curves are sufficient to illustrate the trend. Note also that a graph is better unboxed if the ticks do not extend around all four sides.

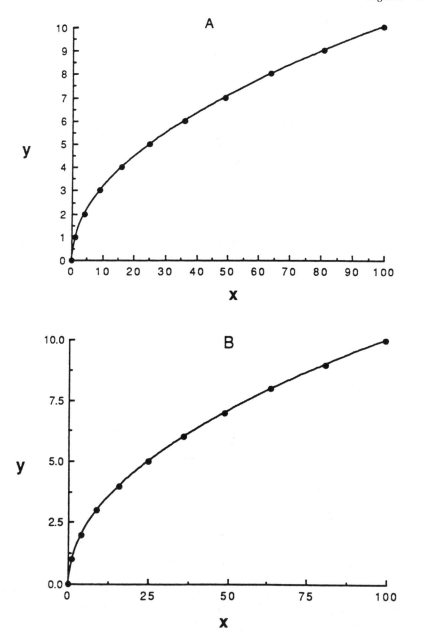

Fig. 5. The scales on the axes: *(A)* Overcrowded, and *(B)* improved.

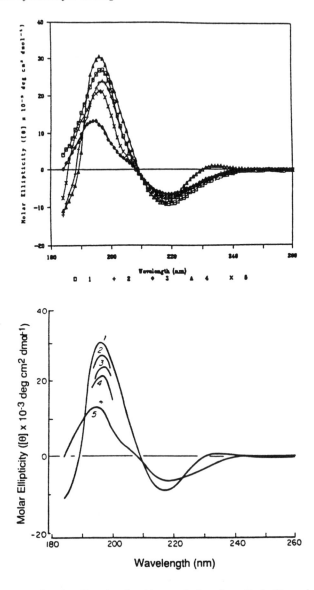

Fig. 6. CD spectra of porins. Reprinted, with permission, from Park, Perczel, and Fasman, "Differentiation between transmembrane helices and peripheral helices by the deconvolution of circular dichroism spectra of membrane proteins," (1992) *Protein Sci.* **1**, 1032–1049. © 1992 by Cambridge University Press, Cambridge, England.

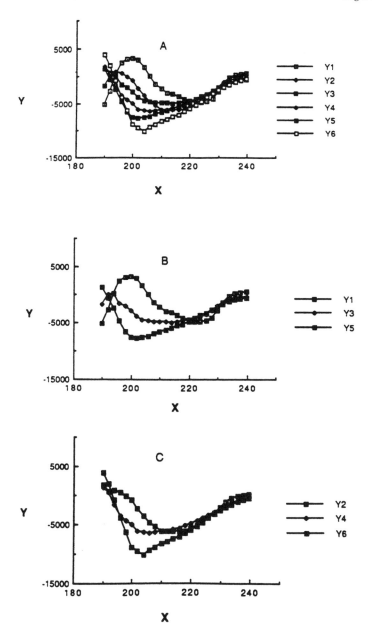

Fig. 7. Separation of a series of six curves *(A)* into two sets of three curves each *(B and C)*.

13. Keep the graph simple.

Ideally, each graph presents one idea and should contain no more than three or four curves. Pertinent information buried in an overcrowded figure may escape the reader's attention. For clarity separate one crowded graph into two or more simpler figures. Figure 7A illustrates a series of six curves that are better plotted as two graphs (Figs. 7B and 7C).

14. Some graphs require different scales in different regions.

For example, in Figure 8, the scale of the ordinate in the far-UV region (below 240 nm) is 100 times that in the near-UV region (240–320 nm). If the same scale were used over the entire graph, the detail above 240 nm would be lost as the curves became virtually flat.

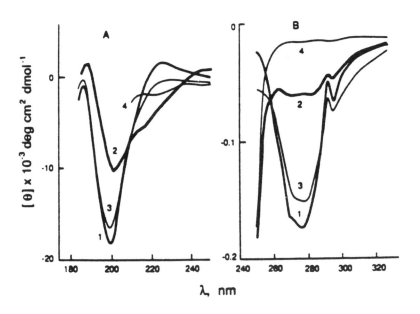

Fig. 8. Mean residue ellipticities of soybean trypsin inhibitor in the *(A)* far-UV, and *(B)* near-UV regions. Curves: *1,* in water at 23°C; *2,* in water at 80°C; *3,* same as 2, but cooled down to 23°C; *4,* in 6 M guanidine hydrochloride plus a reducing agent. Reprinted, with permission, from Wu, Yang, and Wu (1992) *Anal. Biochem.* **200,** 359–364. © 1992 by Academic Press, Inc., Orlando, FL.

15. Display similar curves as a series of graphs.

Similar curves that would overlap should be separated. To avoid overlap, the three curves in Figure 9 are displayed as a series of graphs, with the ordinate scales plotted alternately on the left and right sides for clarity.

Conversely, two or more physical properties can be combined into a single graph, as illustrated in Figure 10. Remember that journals always try to save space, and some limit the number of figures.

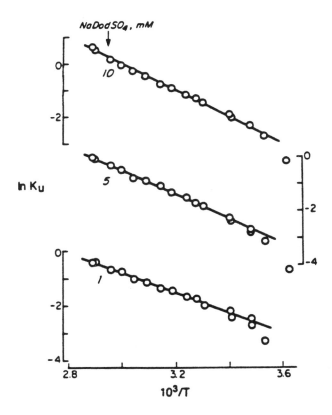

Fig. 9. van't Hoff plots of the succinyl C-peptide analogue of ribonuclease A in $NaDodSO_4$ solutions. Reprinted, with permission, from Wu and Yang (1990) *Biopolymers* **30**, 381–388. © 1990 by John Wiley and Sons, New York, NY.

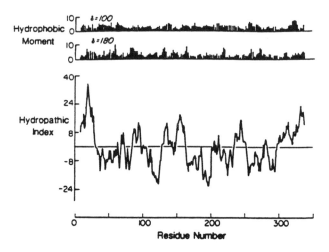

Fig. 10. *Top,* Hydrophobic dipole moments, and *bottom,* hydropathic indices along the amino acid sequence of opioid-binding cell adhesion molecule. Reprinted, with permission, from Wu, Hasegawa, Smith, Loh, Lee, and Yang (1990) *J. Protein Chem.* **9**, 3–7. © 1990 by Plenum Publishing Corp., New York, NY.

16. Use vertical bars to indicate the ranges of experimental error.

Figure 11 shows a standard for calibrating circular-dichroism instruments. The two I-bars represent the ranges of error at the maximum and minimum points. The larger bar indicates that variations were much larger at lower wavelengths which approached the instrument limitation.

17. Do not extrapolate a curve beyond the actual data unless justifiable.

The broken line in Figure 11 indicates that the extrapolated band is Gaussian (symmetrical and bell-shaped), although there are no experimental data to verify this assumption.

18. Identify the source of an illustration that has been published elsewhere.

It is essential to obtain written permission from the copyright holder; this fact should be included in the credit. Note that the copyright holder is not necessarily the author of the illustration.

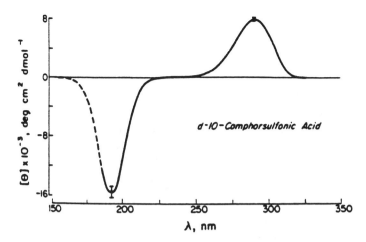

Fig. 11. Circular dichroic spectrum of *d*-10-camphorsulfonic acid in water. The vertical bars represent variations of ± 1.5 and $\pm 5\%$, respectively. The broken line represents the extrapolation of a Gaussian band. From Chen and Yang (1977) *Anal. Lett.* **10**, 1195–1207. Marcel Dekker, Inc., New York, NY. Reprinted by permission of the publisher.

19. Ask yourself: Is the figure necessary?

Your data may be described equally well in the text. For example, at constant pressure P, the volume V of a gas has a linear relationship to temperature T. This relationship could be illustrated in a graph as a straight line, or it could be described in the text as $PV = nRT$, where n is the moles of gas and R is a constant.

Sometimes there is reason to plot a straight line, such as when the linear relationship is confined to a narrow region, beyond which it is no longer straight and extrapolation is unwarranted. Another case is the double-reciprocal plots of the enzyme kinetic data for acetylcholinesterase in the presence of the inhibitor tacrine. This produces a family of straight lines with a common intercept on the abscissa (Figure 12), which in enzymology indicates noncompetitive inhibition.

Finally, if the experimental points scatter, the reader should see the actual plot of the data, which may not be best represented by a straight line.

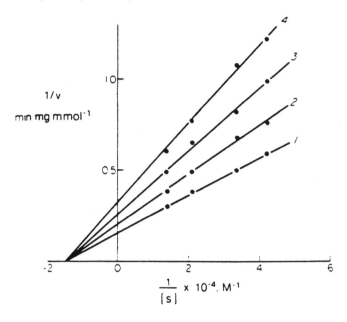

Fig. 12. Double-reciprocal plots of the enzymic activity of acetylcholinesterase versus substrate concentration in the presence of the inhibitor tacrine. Concentrations of tacrine: *1,* zero; *2,* 4 nM; *3,* 8 nM; *4,* 12 nM. Reprinted, with permission, from Wu and Yang (1989), "Tacrine Protection of Acetylcholinesterase from Inactivation by Diisopropylfluorophosphate: A Circular Dichroism Study," *Mol. Pharmacol.* **335**, 85–92. © 1989 by Williams & Wilkins, Baltimore, MD.

B. LEGENDS

A figure becomes complete and intelligible only when an explanatory legend is attached. A legend is needed to explain all symbols, identify all curves, and define all abbreviations not specified on the curves. (Note that most journals now prohibit abbreviations along the curves and allow only numerals.) The legend may also briefly mention experimental conditions such as buffer, pH and temperature. Symbols and abbreviations that are used for similar curves in several figures need only be defined in the figure in which they first appear and cross-referenced in subsequent figures.

Almost all publications now allow abbreviations to be defined in a footnote to the text. It is best to avoid footnotes for abbreviations used in legends, because the reader often glances over the figures before reading the text.

Some journals allow a figure to contain only a legend, which must be brief. Others allow a short title, followed by a descriptive sentence or brief paragraph. A few journals even permit experimental procedures to be detailed in the legend. A journal that imposes a limit to the number of words often excludes legends from the count; thus a descriptive legend can free up space in the text. Each journal's policy can be found in its *Instructions to Authors*.

12 Tables

Experimental data are presented only once—in the text, in a table, or in a figure. A table can list large amounts of numerical values in a small space, and is superior to a lengthy explanation in the text when data are extensive. It is preferable to figures when exact values are of utmost importance.

A table is less effective than a figure for showing trends in the data. A notable example is the binding of oxygen to hemoglobin. If the data are plotted against pressure, the saturation curve is sigmoidal, suggesting that hemoglobin is allosteric (Figure 13). That is, the binding of oxygen to hemoglobin is cooperative and the initial binding enhances additional binding

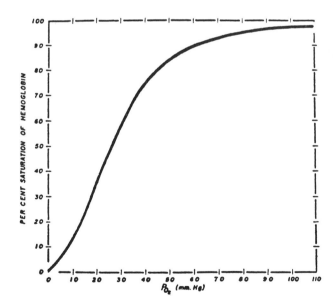

Fig. 13. The binding of oxygen by hemoglobin. Computer plot by I. Sato.

in the same protein molecule. This trend is obvious from the figure, but would not be from a table of data.

A. TABLE COMPONENTS

A standard table, such as Table 1, consists of the following components:

Table 1   Mean residue ellipticities of four respresentative proteins
($[\theta]$x10^{-3} deg cm $\leq$ /dmol)

λ nm	Myoglobin	Elastase	Lysozyme	Flavodoxin
240	-2.29	-0.42	-1.40	-1.65
235	-7.14	-0.84	-3.91	-3.14
230	-15.70	-0.92	-6.14	-5.28
225	-23.80	-1.76	-7.12	-8.24
220	-24.40	-3.14	-8.11	-9.90
215	-21.70	-4.83	-9.43	-9.24
210	-22.80	-7.21	-9.51	-7.92
205	-14.00	-8.89	-0.53	-3.63
200	16.20	-12.40	12.60	6.93
195	51.10	-9.34	12.80	13.50
190	52.90	-2.10	5.78	9.90

Note: Myoglobin, elastase, lysozyme, and flavodoxin represent all-α, all-*fl*, α+*fl*, and α/*fl* proteins, respectively

Table number. The table number is usually Arabic, but may be Roman in some journals. It is numbered in order of appearance in the text. The numbering starts over with each appendix; for example, Tables in Appendix A would be numbered A-1, A-2, and so forth. The table number appears at the top of the table, centered or flush left, and is not terminated with a period.

Title. The title should be a brief and intelligible identification of the contents. Do not include background information, description of results, interpretation, or comments; these belong in other sections of the paper. For example,

Table 5 Binding isotherm of oxygen by hemoglobin at 10, 20, ..., 100 mm
Hg of oxygen pressure

should be shortened to

Binding of oxygen by hemoglobin

A table title may have a subheading, which appears on a new line and
is usually enclosed in parentheses. A typical subheading could be the units
of measure used in the table.

The title and its subheading can be either flush left or centered, and
either follow the table number on the same line, or begin on the next line.

Column headings. A column heading briefly indicates the nature of the
data in the column. It may additionally carry a subheading on a new line,
optionally enclosed in parentheses, which often indicates the units of
measure for the column.

Horizontal rules. Horizontal lines, or *rules*, are drawn above and below
the column headings (below the subheadings, if any). If there is more than
one level of data in the table, this is indicated by a heading for each level
and a horizontal rule spanning the columns in that level. Another horizontal
rule delineates the bottom of the body of data.

Stub. The leftmost column is called the *stub*, and lists categories or
subjects described in the other columns. That is, it indicates the contents of
each horizontal row. The stub requires a column heading only if the contents
are not clear from the table title.

Body of table. Numerical data presented in the table should be decimal-
aligned. Numbers without decimal digits are aligned along the implied
decimal point.

0
12.5
100.7

A missing datum should be indicated by an em dash (—), two hyphens, or
three or more spaced periods (...). Sections for which the categories are not

applicable should be left blank; *N.A.* (for Not Applicable) is sometimes used but is unnecessary.

Footnotes. Footnotes appear at the bottom of the table, flush left. Each footnote begins on a new line; an extra blank line is added between notes for readability. Some journals avoid using letters and numbers (which can be mistaken for exponents) as footnote symbols, and instead use the typographical symbols * (asterisk), † (dagger), ‡ (double dagger), § (section mark), ‖ (parallel lines), and # (number sign).

B. DESIGN CONSIDERATIONS

Although tables will be reset by the journal according to the house format, keep the following in mind when preparing a table:

1. Design the table to fit the columns of the journal. Most journals have two columns per page, some have one column, and a few have three columns.

2. Keep the table simple and easy to follow.

3. Do not waste large amounts of empty space in the body of a table. Large amounts of space in the data section usually indicate poor design.

4. Round the data to the nearest significant figures. Most experiments give data good to at most three digits, whereas computer readouts usually exceed three digits. Using your data with more than three digits would give a false impression of extra precision.

5. Arrange the columns to facilitate comparison of the data. Comparisons between adjacent vertical columns are easier than between horizontal rows.

6. Ask yourself: Is the table necessary? Data that can be described equally well in the text should not be tabulated. For example, data that can be graphed as a straight line with slope m and intercept b on the ordinate can be described simply as $y = mx + b$.

The personal computer has made the creation of illustrations much easier. Professional-looking figures, graphs, diagrams and tables can be produced on a computer quickly and inexpensively by using graphing software such as Cricket III or Kaleidagraph for Macintosh. Figures can be scaled to size and table columns adjusted as needed. Data can even be converted from one type of display into another (line graph to bar graph, for example) as often as you wish. With a variety of font styles and a laser printer, the quality of a software-produced illustration is as good as a photo print.

EXAMPLES.

Table 2 is from my thesis in 1952. Note that its title lacks definitions for the abbreviations A and SDBS, and its subheading would be better placed in a footnote. The column headings take up five lines and there are no horizontal rules delineating the sections. The surfactant concentrations cover a wide range, and might be better expressed in millimolar quantities rather than in ($\times 10^5$ M).

Table 3 is an improved version of Table 2. Its title names the methods used (binding and electrophoretic analyses) but relegates further explanation to a footnote. The significant figures of the experimental data are limited to one, two, or three, depending on the surfactant concentrations used.

Table 4 is an example of data that can be summarized in the text, which would use less space than the table. This table can be replaced with two sentences:

The X-ray diffraction results of concanavalin A show 2% α-helix, 51% β-sheet, 9% β-turn and 37% unordered form (Reeke et al., 1975). The corresponding CD estimates are 4% α-helix, 53% β-sheet, 9% β-turn and 34% unordered form (Provencher and Glöckner, 1981).

Table 2 Combination of A with SDBS at 1–3° in Phosphate-NaCl Buffer (pH 7.7, μ 0.20). A: 5.9×10^{-5} M. Reprinted from Yang and Foster (1953) *Journal of the American Chemical Society* 75, 5560–5567. Copyright 1953 American Chemical Society, Washington, DC.

Total SDBS concn. $\times 10^5$ M	Free SDBS concn. $\times 10^5$ M	Av. moles bound per mole protein, r	A	AI_m	AI_n	AI_{n+x}
				Relative areas, [1,2]%		
0	0	0	100			
16.7	0.2	2.8		100		
41.8	1.0	6.8		100		
83.6	5.2	12		100		
104.4	7.9	15		87.9	12.1	
293	26.1	41		15.8	84.2	
418	34.6	60				100
882	83.7	120				100
1871	159	260				100

[1]Relative areas were calculated from the descending electrophoretic patterns.
[2]m varies from one to about ten, n is about 48 assuming the maximum m is 12, and x is a variable.

Table 3 Binding and electrophoretic analyses of bovine plasma albumin (A) with sodium dodecylbenzene sulfonate (I)

Total I (mM)	Free I (mM)	Bound I/A (mol/mol)	A (%)	AI_m^* (%)	$AI_n^†$ (%)	$AI_{n+x}^{†‡}$ (%)
0	0	0	100			
0.167	0.002	3		100		
0.418	0.010	7		100		
0.836	0.052	12		100		
1.04	0.079	15		88	12	
2.93	0.261	41		15.8	84.2	
4.18	0.346	60				100
8.82	0.837	120				100
18.7	1.59	260				100

Note: Albumin (0.059 mM) in phosphate-NaCl buffer (pH 7.7, μ 0.20) at 1–3°C. Binding isotherm results were obtained from equilibrium dialyses of the mixtures of albumin and surfactant against buffer and percent complexes calculated from the relative areas of the descending electrophoretic patterns.
*The m values vary from 1 to about 10.
†n is about 48 assuming a maximum m of 12.
‡x is a variable.

Table 4 Secondary structure of concanavalan A

Conformation	X-ray[*]	CD[†]
α-Helix	2	4
β-Sheet	51	53
β-Turn	9	9
Unordered	37	34

[*]Taken from Reeke *et al.* (1975).
[†]Estimated from the method of Provencher and Glöckner (1981).

13 Discussion

The Discussion takes the data reported in the Results section and interprets the findings, evaluates their significance, and examines the implications. This is probably the most challenging section to write and will demonstrate how well you understand the results. This does not mean that the discussion should be lengthy, especially if there is little to discuss. In fact, some journals discourage discussion beyond four or five double-spaced typed pages.

A. DISCUSSION FORMAT

The Discussion moves from specific topics to the general, starting with the question or problem posed in the Introduction section. This question should be answered by a chain of arguments that lead logically from idea to idea, not from finding to finding; the argument should not derive from the findings. Comparison with others' observations and conclusions, and the author's unexpected findings, if any, can be included here.

The beginning and ending of the Discussion section are prominent places reserved for important ideas. End the Discussion with a positive statement rather than equivocations or more questions. That is, finish with a conclusion if possible.

How to present the argument is a matter of personal style. The following suggestions are general guidelines for developing the Discussion:

1. Begin the discussion with a topic sentence that returns to the question raised in the Introduction section.

2. Mention new findings, knowledge, and concepts that resulted from your study. Do not, however, introduce data that were not presented in the Results section.

3. State whether you have achieved your goal of answering the research question or have found exceptions and unexplained results.

4. Compare your results and interpretations with related published work, even though it may disagree with yours. Give due credit to others whose work has been confirmed. Be fair with those whose results differs from yours and explain, if possible, the disagreement impartially.

5. Take care to label speculations as such. Journals permit some reasonable speculations if based on solid evidence.

6. Discuss any theoretical implications and possible applications of your findings.

7. Present the conclusions concisely. If additional experiments are needed to validate your results, be sure to qualify your conclusions.

8. Suggest future studies, if any.

9. End your discussion with a short summary or conclusion.

10. Do not repeat material that was presented in other sections of the paper. This is a common occurrence in papers but should be avoided.

Sometimes the author needs to discuss preliminary results to explain why a subsequent procedure was performed. This can be handled by integrating the Results and Discussion into one section.

B. GRAMMATICAL STYLE

Most journals have similar grammatical conventions for the Discussion section. Verb tenses are used to indicate the timing of an action and its state (i.e., whether the action was completed or is still taking place). They help the reader understand when events occurred, relative to some reference point in time. In scientific papers, the customary time frame of reference is the period from the initiation of your research project through the writing of the Discussion.

The past tense is used to describe past events that took place during the project. These include procedures and observations.

Data were taken from fifty volunteers.

The distribution followed a standard bell-curve.

The present tense is used to describe interpretations, conclusions, and implications, because they represent your current beliefs.

We conclude that regular exercise reduces the severity of osteoporosis.

The past perfect tense is used to indicate that an action was completed prior to the time frame of reference, i.e., before your research project started.

Wu had studied the secondary structure of protein A by the sequence-predictive method in 1985.

Use the present perfect tense to indicate that an action was completed either in the present, or at a time that is unknown or unimportant.

The conformation of protein B has been studied by nmr (Smith, 1990).

14 Acknowledgments

The Acknowledgment section is used to give credit to those who have materially contributed to the research. Technical assistance, advice from colleagues, and other research-related contributions can be included here, but contributions that do not involve researching (such as clerical assistance, word processing, or encouragement from friends) should not appear in Acknowledgments.

Grants, gifts, fellowships, and other sources of funding are mentioned here. Although these are not research contributions, the reality is that financial support is the lifeblood of scientific research, and failure to acknowledge patrons in a prominent place is ungracious as well as self-defeating. While acknowledgments are usually placed at the end of the text, some journals differentiate financial support by placing such acknowledgments in an unnumbered footnote on the title page.

Note that American spelling is *acknowledgment,* as compared to the British spelling *acknowledgement* (see Chapter 2 for a comparison of American and British styles).

As always, be brief. Such expressions as

The authors thank

We are deeply indebted to

We are grateful to

The authors wish to acknowledge

can simply be replaced by

We thank ....

A simple acknowledgment may be written as,

> We thank Dr. A. B. Chen for her comments on the manuscript and Mr.
> X.Y. Zataki for his technical assistance. This work was supported by U.S.
> Public Health Service Grant GM-10880-32.

Be aware, however, that the reader may not interpret the acknowledgment in the way you intended. In the preceding example, the reader may infer that Mr. Zataki did all the experimental work and Dr. Chen explained the data, while your contribution was limited to being the armchair general who wrote the manuscript. Make certain that the acknowledgments accurately reflect the situation.

Similarly, the mention of a scientist is sometimes construed as giving endorsement of the paper, while in reality the scientist may have only read your manuscript and even completely disagreed with your treatment. For this reason, it is often desirable to obtain prior permission from the person being acknowledged.

15 References

Almost every research project relies in part upon the work of others. Authors are required to identify their sources of information, that is, to cite references, not only to give credit where it is due, but also to provide the reader with access to these sources.

External material must be identified whenever it is used in the text, but to insert documentation at each occurrence would be distracting and cumbersome. The reference must be documented systematically and unobtrusively.

The two basic components of a documentation system are the *text citation* and the *reference list*. The citation is a brief identification of the information source, and appears in the text somewhere within the paragraph where the information is used. A full bibliographic version of the citation appears with similar listings in a separate reference list, usually following the text. Bear in mind that a reader desiring more information may wish to access some of your references; therefore, the reference list must provide all bibliographic information necessary to identify and locate each source.

There are several documentation systems, three of which are described in this chapter, and many minor variations are acceptable as well. Additionally, style conventions for the reference list vary from one journal to the next. Refer to the journal's *Instructions to Authors* and use their preferred style. The following illustrates a few of the styles in current use:

Chou, P. and Fasman, G. D. (1974) *Biochemistry* **13**, 222–245.

Kubota, S. & Yang, J. T. (1984) *Proc. Nat. Acad. Sci. USA* **81**, 3283–3286.

Creamer, T. P. and G. D. Rose. 1995. Interactions between hydrophobic side chains within α-helices. *Protein Sci.* 4:1305–1314.

Manavalan, P. & Johnson, W. C. Jr. *Nature* **305**, 831 (1983).

A. C. de Dios, J. G. Pearson, E. Oldfield *Science* **260**, 1491 (1993).

A. NUMBER SYSTEM

Under the number system of documentation, the reference list is arranged in order of the first citation of each source in the text. The sources are numbered sequentially, starting with 1, and are cited in the text by number. The citation appears in parentheses, brackets, or superscript; for example, (1), [2–5], or [6].

The history of optical activity goes back almost two centuries to the pioneer researches of J. B. Biot and A. J. Fresnel (1). Biot was the first to observe two types of optical rotatory dispersion. He also gave us the definition of specific rotation (2).

1. Smith, J. Q. (1978) *Famous French Scientists.*
2. Lowry, A. (1935) *Optical Rotatory Power.* Longmans, Green, London; (1964) Dover, New York.

Bovine serum albumin in acid solution undergoes reversible "swelling" upon lowering the *p*H from 4 to 2.[1]

1. Yang, J. T. and Foster, J. F. (1954) *J. Am. Chem. Soc.* **76**, 1588–1595.

A numerical citation is brief and unobtrusive. This is particularly advantageous in scientific papers, where several references may be given in one citation. On the other hand, the reader must turn to the reference list to identify the work, which is apt to be disruptive. Also, additions or deletions require that both the text citations and reference list be renumbered.

B. AUTHOR-DATE

The author-date system of documentation is also known as the Harvard system. The reference list is arranged alphabetically by the name of the first author and the date of publication. The basic text citation consists of the last names of the authors and the year of publication; most journals separate author and date with a comma. Under this system the preceding citations would be written:

> The history of optical activity goes back almost two centuries to the pioneer researches of J. B. Biot and A. J. Fresnel (Smith, 1978). Biot was the first to observe two types of optical rotatory dispersion. He also gave us the definition of specific rotation (Lowry, 1935).

> Bovine serum albumin in acid solution undergoes reversible "swelling" upon lowering the pH from 4 to 2 (Yang and Foster, 1954).

The chief advantage of the author-date system is that the citation provides the reader with more bibliographic information than the number system. Sources can be easily added or deleted to the reference list up to the time that the manuscript is typeset, without changing the existing citations and references. One potential drawback to the author-date system is that a citation containing several references takes too much space and is visually distracting to the reader.

NUMBER SYSTEM

Many methods of circular dichroic analysis have been developed (1–19) and all use the far-UV spectra of a set of reference proteins to determine the conformation of an unknown protein.

AUTHOR-DATE SYSTEM

Many methods of circular dichroic analysis have been developed (Greenfield and Fasman, 1969; Rosenkranz and Scholten, 1971; Brahms and Brahms, 1980; Saxena and Wetlaufer, 1971; Chen and Yang, 1971; Chen et al., 1972, 1974; Chang et al., 1978; Bolotina et al., 1980a, 1980b;

> Provencher and Glöckner, 1981;
> Hennessey and Johnson, 1980;
> Compton and Johnson, 1986;
> Manavalan and Johnson, 1987;
> van Stokkum et al., 1990;
> Perczel et al., 1991, 1992;
> Böhm et al., 1992; Sreerama and
> Woody, 1993) and all use the
> far-UV spectra of a set of
> reference proteins to determine
> the conformation of an unknown
> protein.

In this case the number system is clearly preferable to the author-date system, which requires the 19 references to be listed in the text. One possible compromise if using the author-date system is to present a particularly long citation as a footnote, thereby providing the same information nearby with minimal disruption.

Many methods of circular dichroic analysis have been developed[1] and all use the far-UV spectra of a set of reference proteins to determine the conformation of an unknown protein.

---

[1]Greenfield and Fasman, 1969; Rosenkranz and Scholten, 1971; Brahms and Brahms, 1980; Saxena and Wetlaufer, 1971; Chen and Yang, 1971; Chen et al., 1972, 1974; Chang et al., 1978; Bolotina et al., 1980a, 1980b; Provencher and Glöckner, 1981; Hennessey and Johnson, 1980; Compton and Johnson, 1986; Manavalan and Johnson, 1987; van Stokkum et al., 1990; Perczel et al., 1991, 1992; Böhm et al., 1992; Sreerama and Woody, 1993.

The key points of the author-date system follow. Note that journals often reflect minor variations from the conventions described below. As always, the particular journal's style takes precedence over other conventions.

1. Text Citation

In its simplest form the text citation combines the last name of the authors with the year the work was published, and the citation is placed within parentheses. In most scientific publications the author and date are separated by a comma.

(Yang, 1994)

(Samejima and Yang, 1969)

If the date, as given in the reference list, has a letter appended, both year and letter are cited.

(Gray, 1987b)

The date by itself is sufficient when the author is mentioned in the same sentence as the citation.

Chou and Fasman (1974) proposed a sequence-predictive method for determining the secondary structures of proteins.

When the reference list contains more than one author with the same last name and publication year, the initials of the authors' given names can be included in the citation.

(Lee, Y., 1984)

(Lee, M., 1984)

If the citation refers to a portion of the work, it may be helpful to specify this.

(Darcy, 1978, 23)

(Nicholas, 1989, vol. 2)

(Gene, 1958, fig. 5)

(Hertz, 1992, 3:24–32)

Note that the abbreviation for *page* or *pages* is omitted, as is *vol.* when the volume is specified with page numbers. Other abbreviations such as *fig.* and *vol.* (without page numbers) should be included.

Several references can be included in one citation, separated by semicolons.

(Gates, 1980; Dale and Sanders, 1992)

If the additional references are by the same authors, the names are not repeated and the dates are separated by commas.

(Chen et al., 1972, 1974)

Multiauthor works. For a work with more than two authors, use the name of the first author followed by *et al.* in roman type, with no preceding comma. Thus, if the authors are Ames, Botts, and Chen, the citation is

(Ames et al., 1975)

The abridgment using *et al.* makes the citation less obtrusive. It may happen, though, that the reference list contains another work that would abbreviate to the same citation. Suppose, for example, that the reference list contains

Adams, A., B. Botts, and C. Chen (1975) *Handbook of Chemistry ....*

Adams, A., W. Wong, and L. Lee (1975) *Methodology Review ...*

Under the author-date system both works would be cited as

(Adams et al., 1975)

To avoid ambiguity, differentiating information must be added to both citations: either the full author lists, or shortened titles. For example,

(Adams et al., *Handbook,* 1975)

(Adams, Wong, and Lee, 1975)

Abbreviating the author list. If the author is an organization, the name may be abbreviated in the text citation, although the full name and abbreviation must be provided in the reference list.

(NIH, 1974)

National Institutes of Health (NIH) (1974) A report on legionnaire's disease. *J. Am. Med. Assoc.* **102**, 232–248.

Date of publication. A work that is assured of publication but not yet released is designated *in press;* the publication date can be cited if known.

(Smith, in press)
or
(Smith, 1995)

2. Reference List

The reference list appears after the text. Each entry begins with the author and date, and the entries are arranged alphabetically by author and date to facilitate cross-reference with the text citations.

Author list. The first author's name is inverted (that is, the family name appears first) to simplify alphabetizing, and given names are usually abbreviated to initials separated by a space. Some journals invert the names of all the authors. If the author is an organization, the name is not inverted and leading articles (such as A or The) are omitted.

Smithsonian Institution

Yang, J. T., Chen, G. C., and Jirgenson, B.
or
Yang, J. T., G. C. Chen, and B. Jirgenson

Formerly, all authors of a work were included in the reference listing. These days, however, it is not uncommon for a publication to have ten or more authors (the record may be over 200 authors; see Chapter 6 Authors). Some journals now specify that for a work with more than three authors, only first author be named, the rest being replace by *et al.* in roman type.

A 3-em dash (six hyphens may be substituted) can replace the names in successive works by the same group of authors or editors, although many journals repeat the author list.

Kuzo, T. and B. Franks (1990)
------ (1991)

or

Kuzo, T. and B. Franks (1990)
Kuzo, T. and B. Franks (1991)

The 3-em dash is not used if the order of authors or editors changes.

Date of publication. The publication date follows the author list. Most scientific journals enclose the date with parentheses (although for other scholarly works, many publishers omit parentheses and punctuate the date with a period).

The reference list may contain more than one work from the same author or group of authors in the same year. These are differentiated by appending a roman letter to each year, in alphabetical sequence.

Gray, G. (1987a)

Gray, G. (1987b)

A work that is in the process of publication is documented in the same manner as a published work, except that the publication date is replaced by *in press*. The publication date can be used if known.

Title. The date of publication is often followed by the title of the work, although if the work is a journal article, the title is usually omitted. The title of a chapter or article, if used, is given in roman type without quotation marks and punctuated with a period. It is capitalized sentence style; that is, only the first word and proper nouns are capitalized.

This is followed by the title of journal, periodical, or book, which is given in italics, without quotation marks. The original capitalization is retained. To save space, journal titles are abbreviated; it can be assumed that scientists are sufficiently familiar with the titles to recognize them in abbreviated form. For consistency and clarity, abbreviate journal titles as listed in Chemical Abstracts Service Source Index (CASSI) or *International Serial Catalogue*. Serial publications such as *Advances in Protein Chemistry, Annual Review of Biochemistry,* and *Methods in Enzymology* are abbreviated as journals to *Adv. Protein Chem., Annu. Rev. Biochem.,* and *Methods Enzymol.,* respectively. When in doubt, spell the title in full.

Many books have different authors for the foreword, introduction, and individual chapters. Specific contributions are differentiated from those of the general author or editor.

DeVogelaere, René (1978) Foreword to *Disquisitiones Arithmeticae,* by K. F. Gauss.

Yang, J. T., Chen, G. C. and Jirgenson, B. (1976) In *Handbook of Biochemistry and Molecular Biology,* 3rd ed., Fasman, G. D., ed.

A work that has not been accepted for publication is usually unavailable to the public, so it is excluded from the reference list. The text citations for such works parenthetically state the publication status, such as *manuscript in preparation, manuscript submitted, unpublished data,* or *personal communication.*

Denaturation of protein A at pH 13 was irreversible (unpublished data).

An unpublished dissertation or thesis is usually available to the public and is therefore included in the reference list if the publisher permits, although most scientific journals do not. If included, documentation is similar to that of a published work, except that the work is identified as a dissertation or thesis, the title is enclosed in quotation marks, and the name of the academic institution replaces that of the publisher.

Anderson, K. G. (1995) "Genetic Engineering: Tools for the Next Century." Ph.D. diss., Harvard University.

Volume and issue. The title of the journal is followed by the volume number, which is given in Arabic numerals. It is not necessary to include the word *Vol.* unless the volume is likely to be mistaken as a page number. Most scientific journals use bold-face type for the *Vol.* and number, to further avoid confusion.

The issue number, if any, follows and is separated from the volume by a comma and the abbreviation *no.* Alternatively, the issue can be enclosed in parentheses. Whichever style you choose must be employed consistently throughout the paper.

Gabriel, K. (1995) *Laboratory Technique* 6, no. 2
or
Gabriel, K. (1995) *Laboratory Technique* **6** (2)

Page numbers. Page numbers are usually necessary for a journal article, since the title is often omitted. A single page is indicated by a single page number, while a range of pages is indicated by the first and last page numbers, separated by an en-dash or hyphen. (A en-dash resembles a hyphen but is slightly wider. It is used to indicate ranges.) These elements are separated by commas.

100, 105, 110–112

There are several ways to indicate pages in the reference list. One method is to precede the page numbers with a colon and a space. Another is to precede the page numbers with *p.* or *pp.* If, as is frequently the case, the volume and issue are bolded, a comma before the page numbers is sufficient.

Chou, P. and Fasman, G. D. (1974). *Biochemistry* 13:222-245.

Chou, P. and Fasman, G. D. (1974). *Biochemistry* **13**, 222-245.

Yang, J. T., Chen, G. C. and Jirgenson, B. (1976) In *Handbook of Biochemistry and Molecular Biology,* 3rd ed., Fasman, G. D., ed., *Proteins,* **Vol. 3**, CRC Press, Cleveland, pp. 3-140.

Publication facts. The name and place of the publisher is provided to help the reader obtain the work. These facts need not be provided for most journals and periodicals, whose wide distribution makes them easy to obtain. Exceptions are some foreign journals, smaller publishing houses, and journals with common names that might be mistaken for others.

The reader must be able to locate the publisher's main office. For larger publishing houses, it is usually sufficient to provide the name of the city. Otherwise, provide more information—even a complete mailing address if necessary.

C. ALPHABET-NUMBER SYSTEM

The alphabet-number system is identical to the number system, except that sources in the reference list are arranged alphabetically by author. It has the added advantage that the alphabet arrangement makes it easier for the reader to locate a reference. This system is still not as widely used as the number and author-date systems, however.

It is the author's responsibility to list the references accurately. This may seem an obvious point, but many reference lists are full of typographical mistakes and errors of context. Text citations must agree with the corresponding reference list entries, and all bibliographical information must be correct. Do not rely on memory for bibliographical data, not even for your own publications; authors have been known to err in reporting their own coauthors, page numbers, and even journal names.

Julius Comroe had a good sense of humor and often used amusing anecdotes in his lectures to illustrate a point. Speaking on the need to verify the accuracy of references, he excerpted a letter to *Science* by a Dr. D.C. Blanchard:

A while ago I had occasion to reference a paper ... in *Tellus* ... [I]t was cited by the authors of six or seven papers in the 21 September 1972 issue of the Journal of Geophysical Research ...[The first author] said that it had appeared in volume 21, 1970, pages 451–461 ... [T]he second author said the paper appeared, not in volume 21, but in volume 22 ... [A] third author said the second author was correct regarding the volume number, but that both were wrong about the page spread. The paper actually ends on page 462 ... Fast losing faith, I turned to a fourth author ... [He] was none other than the author of the *Tellus* paper in question ... I assumed that he would cite his own paper in impeccable form. He said everyone was wrong—the paper starts on page 541, not page 451 ... I found a fifth author who said the proper *Tellus* volume was neither 21 nor 22, but 12 ... I went to the last remaining author who used the reference ... [H]e agreed with the second author. The proper volume is 22, and the page spread is 451–461. Authors 2 and 6 are right. I confirmed this at the library.

Another anecdote on the subject of accurate references was called "Dr. Uplavici":

Dr. Uplavici

In 1887 Jaroslav Hlava, distinguished professor of pathological anatomy at Prague, reported his discovery that diarrhea caused by *Entamoeba histolytica* could be transmitted from manto cats. Professor Hlava's paper, written in the Czech language, was entitled "O úplavici. Predbezné Sdeleni," meaning, "On Dysentery: Preliminary communication." In the same year a Dr. S. Kartulis, of Alexandria, reviewed Professor Hlava's report in German and somehow attributed the work to "Uplavici, O." and even referred to correspondence with "Uplavici" in Prague. For the ensuing 50 years investigators and reviewers around the world referred to the classic experiments on amebic dysentery by "Dr. O. Uplavici." Finally, in 1938, "Uplavici" was exposed and laid to rest by Clifford Dobell in London.

16 Summary of Preparing a Manuscript

The motto of scientific writing is brevity and clarity, that is, to provide maximum information with minimum words in a well-organized manner. This quick reference may be helpful during the preparation of a manuscript.

A. FRONT MATTER

The front matter precedes the text and pertains more to the bibliographical facts of the paper than to the actual research. Front matter consists of the title page, the abstract, and a list of key words for indexing.

1. Title

- Choose a title that will attract the reader's interest.
- Use the fewest possible words to adequately describe the content of the paper.
- Be specific.
- Avoid abbreviations, except standard ones such as DNA.
- Put important terms at the beginning of the title

2. Authors and Affiliations

- Include only those who have contributed materially to the research project.
- List order depends on each author's role and contribution.
- Write names in western format: Given names followed by family name. Possible exceptions are names of famous people, which should be given in their most recognizable form. This rarely happens in a scientific paper, however.
- List the affiliations of all authors.

- List the corresponding author, the address to which correspondence should be directed, telephone number, facsimile (fax) number, and electronic mail (e-mail) address, if any. Provide the country code for telephone and fax numbers outside of the United States

3. Abstract

- An abstract is usually less than one double-spaced typed page. It begins on a new page and contains up to 150 to 250 words. If possible, avoid citing references in the abstract.
- Objective and scope: informative for research papers and indicative for reviews, conference reports, and so forth
- Methodology: brief unless the research project is about methods
- Summary of results
- Conclusions

4. Key Words

- Provide several words or phrases for the benefit of the indexer.
- Include words that are not part of the title of the paper.

B. TEXT

The text is the main body of the paper. The organization of the text is represented by the acronym IMRAD, which stands for:

Introduction:What problem was the research project addressing?
Materials and Methods:How did you study the problem?
Results:What did you find?
And
Discussion:What do these findings mean?

1. Introduction

The introduction contains information that should be read before the rest of the text. Its purpose is to provide the educated reader with specifics needed to understand the paper. The introduction typically includes:

- Nature and scope of the problem
- Pertinent literature cited
- Methods
- Recent findings and theories
- Principal results

2. Materials and Methods (Experimental Procedures)

This section describes and justifies your approach to the research problem.

- Provide detail sufficient to enable a competent reader to repeat the experiments.
- Do not include results, except in a methodology paper, in which the methods become the results.
- Remember that a good reviewer will read this section to judge the validity of your approach.

3. Results

This section is the meat of a paper, presenting the findings in text, illustrations, and tables. This does not, however, mean that the Results section should be lengthy.

- Do not start the Results section by describing methodology.
- Report significant results only.
- Avoid redundancy: Numerical values that are apparent in illustrations and tables should not be duplicated in the text.
- Cite figures and tables concisely: Do not include the same data in both figures and tables and do not repeat the legend of a figure or the title of a table in the text.
- Bear in mind that text in figures and tables must be legible after reduction by the printer, and some exhibits are costly to reproduce.

4. Discussion

This is usually the most challenging section to write. It is not a recapitulation of results. The value of the discussion lies in your interpretation of the findings and their significance.

- New results should not be introduced in this section
- Present the principles, relationships, and generalizations shown by the results
- Point out any exceptions or any lack of correlation, and define unsettled points, but avoid focusing on trivial details
- Show how your results and interpretations agree or disagree with published work
- Discuss the theoretical implications and any possible applications of your findings and interpretations
- State your conclusions
- Summarize your evidence for each conclusion

C. BACK MATTER

The back matter follows the text and lists resources that were not part of the research project but nonetheless contributed to its execution. These include research contributions, sources of funding, and reference materials.

1. Acknowledgment

The main purpose of this section is to credit those who have made significant research contributions to your project. Another important function of this section is to mention individuals and entities that have provided essential support such as grants and fellowships.

2. References

- Cite only significant published references
- Follow the journal's instructions for documentation, which will be some version of the author-date system, the number system, or a combination of the two

Part III Publishing a Manuscript

17 Submitting the Manuscript to a Journal

A. PREPARING THE MANUSCRIPT

1. Paper

Use standard size white paper of good quality, preferably bond (office-quality paper) of 20- or 24-pound weight (or 75–90 g/m^2). Standard size paper in the United States measures 8½ × 11 inches; the closest size paper in Asia and Europe is A4, which measures 210 × 297 mm. Do not use "erasable" paper.

2. Typing instructions

Either a typewriter or a computer can be used to type the manuscript. The same type size should be used throughout the manuscript; pica (10 characters per inch) is preferred, but elite (12 characters per inch) is acceptable. If using a computerized word processor, choose a standard font such as Courier or Times Roman in 12- or 10-point size. Remember that very few typewriters and printers can reproduce all of the characters used in scientific writing. Such symbols must be inserted in the manuscript by hand, written in ink.

Use one-inch margins on all fours sides of the paper. All of the material must be double-spaced or triple-spaced; no part of the manuscript is to be single-spaced. Consult the instructions to authors for formatting instructions, such as the paragraph indenting (usually one-half inch), headings, and justification. Following are the headings used for two papers that were submitted to different journals.

Analytical Biochemistry:
(INTRODUCTION)
MATERIALS AND METHODS

Materials
Methods
RESULTS
CD Spectra of Denatured Proteins
Analysis of Unfolding Proteins
DISCUSSION

Journal of Protein Chemistry:
1. INTRODUCTION
2. MATERIALS AND METHODS
2.1. Materials
2.2. Preparation of Solutions
2.3. Circular Dichroism
3. RESULTS
3.1. Protein Conformation in Buffer
3.2. Conformation in Organic Solvents
4. DISCUSSION

3. Assembling sections

The manuscript is usually assembled in the following order:
Title page
Footnotes (Some journals place these after the text)
Abstract and key words
Introduction
Materials and Methods (Experimental Procedures)
Results
Discussion (or Results and Discussion combined)
Acknowledgments
References
Tables
Figure legends
Figures.

Each section begins on a new page, and each table and figure is on a separate page. All pages should be numbered consecutively at the bottom center, starting with the title page. If using a word processor to develop the manuscript, use its automatic page numbering function, which is easier and more accurate than manual numbering.

B. SUBMISSION TO THE JOURNAL

The basic steps for submitting a regular manuscript for publication are:

1. Read the Instructions to Authors of the journal for submitting the manuscript. Note the journal's formatting conventions, which must be employed in your manuscript.

2. List on the title page the title, names of the authors and their institutional affiliations, the running title, and title footnote. The first author listed on the cover page is assumed to be the corresponding author, who acts as the primary liaison with the publisher, although another author can be designated. Furnish the corresponding author's name, mailing address, telephone number, facsimile (fax) number, and e-mail address.

Dr. L. Peller
CVRI, Box 0542
University of California
San Francisco, CA 94143-0542
U.S.A.
Telephone: 01-1-415-476-2891
Fax: 01-1-415-476-2283
E-mail: peller@cvri.ucsf.edu

Naturally, the country name and international dialing code are not necessary for domestic addresses.

3. Type the manuscript, incorporating the journal's formatting. The manuscript must be double-spaced (some journals also accept triple-spacing). If the journal does not express a preference for fully-justified text (that is, even left and right margins), you can use left justification (ragged right margins). Full justification gives a look of neatness and symmetry but creates excess white space in the text. Left justification creates even spacing and makes the text more readable, which may be more important than symmetry.

4. Use of a word processor is highly recommended to facilitate the revision process. If you invest the time to learn its advanced features (also recommended), page and chapter numbering and spelling corrections can be handled automatically. **Be sure to make frequent backup copies of your work.** One caution: A word processor is only as accurate as the person using it. Do not assume that the word-processed text is free of errors. It is your responsibility to double-check the printout.

5. Type reference list, tables, and figure legends. Begin each of these on a new page. The printer will scale the figures to size and add the legends afterwards.

6. Prepare glossy photographic prints or laser prints of the figures. On the back of each print, write the authors' names and the figure number, and indicate the top edge of the figure. Use light pressure when writing, to avoid marring the print.

7. Proofread the typed manuscript. (At this point the manuscript is technically a *typescript,* although the distinction is minor. *Manuscript* can still be used.) Most word processors have a spell-checking function which is indispensable, although it will annoyingly flag technical terms as misspellings, unless it has a scientific dictionary. Grammar-checking software is also available. Bear in mind that these programs cannot detect errors of context. For example, "The substance was coal," would pass as correct, even if you really intended to write, "The substance was cool." Unambiguous errors such as, "Inthe introduction," would of course be caught.

8. Make the required number of typescript copies for the journal. Also make one copy for each author.

9. Write a cover letter to the editor of the journal, listing the enclosures. If you wish, you may *briefly* introduce yourself and your project. You may recommend several potential reviewers and even request that certain reviewers not be chosen. Some journals place restrictions on this privilege; for instance, some will not allow members of their editorial advisory board to be disqualified from taking part in the final disposition of a manuscript.

10. Mail all necessary materials to the editor. Some journals now require submitting the manuscript electronically, that is, on computer diskette. Include in the mailing

- A cover letter
- One copy of original manuscript (not stapled)
- One set of glossy figures or laser prints, protected by a piece of cardboard of the same size
- The required copies of the manuscript, including duplicated figures (each of these copies is stapled)
- Copies of relevant manuscripts in press or submitted for publication elsewhere, so the reviewer can better evaluate your manuscript. Clearly identify these as supplemental manuscripts for comparison purposes.
- One signed copy of the rights transfer form, if required by the journal.

Accelerated Publication. A manuscript of unusual urgency and significant interest may be submitted as a *Letter to the Editor* or *Communication to the Editor*. The paper is brief but is not to be presented as a note or preliminary report. It is required to contain all the elements of a regular manuscript and satisfy all of the journal's publishing criteria. In addition, its findings must be of immediate significance to current research that merits accelerated publication.

The editor should be advised in the cover letter if you wish the manuscript to be considered for accelerated publication.

18 Review and Decision

A manuscript is usually evaluated by at least two anonymous reviewers (referees). The editor or an associate editor will refer the manuscript to a member of the editorial board, who often asks outside referees for additional opinions before delivering a definitive recommendation. The referring editor then makes the final decision on the manuscript.

Some journals subject the manuscript to a preliminary review by two editors, or an editor and a member of their editorial advisory board. The manuscript will be rejected without comment if it is considered inappropriate.

The decision on a manuscript is usually one of the following:

1. Outright acceptance, which is rare,
2. Outright rejection, usually when both reviewers are very negative about the manuscript or one reviewer has raised serious objections,
3. Request for major or minor revision.

Papers of high quality, especially those with important discoveries, are usually accepted immediately, perhaps with minor revisions. This does not mean that a rejected paper is inferior or unimportant. We all know that *Nature* accepted Watson and Crick's paper on the double-helix of DNA, but H. A. Kreb's paper on the citric acid cycle was not accepted at first by a prestigious journal. In a June 1937 letter the editor of that journal did not precisely reject the paper, but regretted that they could not consider it at present, because the pages were full! The manuscript was returned, "in case Mr. Krebs preferred to submit it for early publication to another periodical." All three scientists later became Nobel laureates.

Much more recently, the 1987 manuscript on novel gene amplification method, called polymerase chain reaction or PCR, was rejected by a leading scientific journal. K. B. Mullis, who with his colleagues originally devised the technique, was also invited to Stockholm a few years later. Obviously, no review process can be infallible, but opportunities for publication in other high-impact journals do exist if the first journal has made a mistake.

An outright rejection is better accepted with grace. More experiments or theoretical work may be needed to strengthen the paper. If you feel strongly that the reviewers were wrong and the rejection was unjustified, you can of course contest the editor's decision. Such a fight may not be worthwhile because the revised paper can later be rejected by the same reviewers and you will have invested additional time and effort for nothing. You may be better advised to resubmit the revised paper to a different journal.

Journal editors usually handle rejections quite diplomatically (although of course there are some intemperate ones, who thankfully are in the minority). They may say that the manuscript is suitable for a more specialized journal, or that the work is not sufficiently interesting to their general readership. This allows the editors to reject a manuscript without commenting on its merit. An editor can be magnanimous, too, by saying that the work appears to be well done, but is nevertheless inappropriate for his journal, although he appreciates the author's interest.

The author can appeal in writing, which will then be referred to a member of the Editorial Advisory Board. Decisions reached with the participation of a board member should be considered final.

Negative comments can be quite frustrating, especially when an anonymous reviewer criticizes the author on nonscientific matters. Once I incorrectly referred to the α-helix model as "Pauling-Corey's model" (instead of "Pauling and Corey's model" or "the Pauling-Corey model"). An anonymous referee seized upon this grammatical error and sarcastically asked, "Who was Pauling-Corey?" I protested to the editor that any high school pupil knows who Pauling is and that no one would mistake Pauling-Corey for the hyphenated name of a Ms. Pauling who married Mr. Corey. The paper was immediately accepted.

Authors with English as a foreign language may occasionally encounter prejudice from a reviewer. For example, if all the authors have Chinese names, an anonymous reviewer might disdainfully suggest that "the paper may be more suitably published in China or Taiwan," implying that Chinese journals have lower standards. Such condescending comments are better laughed off because the author cannot fight a referee hiding behind anonymity. Fortunately, an enlightened editor will usually disregard such unprofessional comments.

Perhaps the most unkind, nonscientific comment that I have ever received was from a reviewer who simply did not like the physical technique that we used. In addition to making several derisive comments, he condemned all of us who use the technique as "the terriers [which] continue to gnaw on an old bone." In my letter to the editor I complained that "such condescending remarks are shocking and offensive." I went on to say,

> Perhaps the reviewer's own "meaty" work is all *filets mignon*, be they well-done or otherwise. He professes his "personal prejudice," but I will not dignify his intemperate language with a rebuttal.

The editor immediately sent me the following reply:

> We wish to extend to you our sincere apologies for the comments made by Referee III. We truly regret the unseemly observances made by this reviewer.

Incidentally, our paper was accepted despite (or perhaps because of) this reviewer's outrageous comments.

Once I submitted a manuscript to a biochemical journal of very high standard. To my surprise, the board member who handled this manuscript accepted the paper without revision, except to correct a typographical error. I have had several other papers which were also accepted without revision. These actually caused some inconvenience for me because I needed to make some minor changes to the first proof, or add an addendum. Once a manuscript is accepted, the editors do not expect further modifications.

Most manuscripts need some modification. The author should be able to judge whether the reviewer's criticisms are minor or major. If minor, the manuscript will very likely be accepted after some revision. Very unfavorable or extensive comments are clues that the revised manuscript will only be reconsidered, probably by the same reviewer. Major rewriting can be as troublesome as an outright rejection and it is sometimes wiser to submit the revised manuscript to another journal for consideration. An author can always find some journal to publish the paper, but since reputation is at stake it is unwise to publish just for the sake of publishing.

Many journals request that the revised manuscripts be returned to the editorial office within sixty days, beyond which they will be considered as new submissions and will undergo a new review process. To submit a revised manuscript the cover letter should include the following:

1. The authors' names, the title of the manuscript, and the manuscript number assigned by the journal
2. The authors' response to the reviewers' comments and criticism, including the page and line number of each comment
3. An explanation of any disagreement between the authors and reviewers to the satisfaction of the editor
4. The word "REVISION" typed or handwritten on the required copies of the revised manuscript
5. A copy of the original manuscript with editing marks (only if major alterations were made). This will help the editor and the reviewers verify how what part of the reviewers' suggestions were incorporated into the revised manuscript.

The editor will decide whether the revised manuscript has satisfactorily responded to all the comments. He may then accept the paper or send it to one or both reviewers for further comments. Sometimes the author must revise the paper a second time before it is accepted for publication.

Page Charges. Many journals now have page charges to partially offset publication costs (*Protein Science* is a notable exception). The *Journal of Biological Chemistry* indicates

> [An] accepted manuscript will be published with the implicit understanding that the author(s) will pay a page charge. Current page charges will be obtained by contacting the JBC office. Under exceptional circumstances when no source of grant or other support exists, the author(s) may apply **at the time of submission** for a grant-in-aid to the Chairman, Publications Committee ...

Payment of the charges is not a condition of publication; neither the editors nor the reviewers know if a grant-in-aid or fee waiver has been requested. For American Chemical Society journals, a request for a page charge waiver form should be addressed to Journals Department, Publications Division, American Chemical Society, Washington, DC.

The cost of reprints is expensive, especially for the first 100 copies. Color photographs are extreme expensive. Some authors duplicate their reprints from the published journals without permission. I do not recommend this practice, which is illegal.

19 First Proof

Once the manuscript is accepted by the journal, the text is set into type by the publisher. This often takes one or two months, or even longer for a few journals. The first *proof* (a sample printing to be checked for errors) is then sent, along with any questions from the editor, to the corresponding author. It is the responsibility of the submitting author to ensure that the proof is corrected and all queries from the editor are answered. This first proof may be a *galley proof,* in which the text is not yet formatted into pages, or it may be a *page proof,* paginated as it is to appear in the journal. The proof should be returned to the publisher within 48 hours.

Ideally, two people not involved with the manuscript should *proofread* (check the proof for errors), with one person reading the manuscript aloud while another follows along on the proof, marking corrections. The authors should not proofread, except as a final check. They are usually too familiar with the manuscript to effectively detect errors; the mind tends to see what it knows "ought" to be there.

Corrections other than printer's errors should be avoided. This is not the time to add new text, which must be approved by the editor and may delay publication. Significant new findings that are published after the manuscript has been accepted can, with the editor's permission, be incorporated as an addendum to the text. Computerized typesetting allows the printer to make adjustments comparatively easily, but this is not license to make nonessential or last-minute changes. The manuscript as accepted by the journal should be considered final.

The publisher will provide a list of proofreader's symbols to be used for marking corrections. Corrections are made on the proof (Figure 14). A mark in the line of type indicates where a correction is to be made, and the corrections are written in the margin next to the line of type, using standard proofreading symbols. Always write in the margins, not above the lines of type. The typesetter searches the margins for changes and would miss corrections written between lines of type.

Fig. 14. Example of an edited proof.

In-line marker	Marginal mark	Explanation
Louis Pasteur. In 1848 L. Pasteur	*bold*	bold-face
discovered molecular dissymmetry which	#	insert
laid the foundation of stereochemistry. he	*ital / cap*	italic type/capitalize
began to systematically study the optical		
activity of sodium tartrate ammonium adn	*tr / tr*	transpose
solved the puzzle of why some crystals of		
[this compound were reported to be	[	move to left
chiroptical and others not .)		
⌐ Under his scrutiny, PASTEUR found	*no ¶/lc*	no paragraph/ lower case
that the so-called optically inactive crystals		after initial capital
were not all identical Fig. 1. He separated	(/)	parentheses
the two classes of crystals by hand and		
redissolved them in H2O separately. Both	*m / 2*	substitute/subscript
solutions were chiroptical but of opposite		
sign, and a mixture of equaaal weights of the	⌒	delete and close up
two types of crystals was optically inactive,Y		superscript
] Incidentally, the dextro- and levorotatory salts	⌐	move to right
unite to form a recemate at temperatures		
above 26ØC. Thus we might not have	*Ø / ϡ*	superscript/comma
Pasteurs epoch making discovery had he	*ʾ / =*	apostrophe/hyphen
been living in a climate	*Warmer / ⊙*	insert/period

If a line requires more than one change, the margin corrections are separated by vertical lines and are given in the same order they are to appear in the text. For long corrections or many errors in a small section of text, it is better to cross out the section and type the corrected version on a separate sheet of paper. Label the corrected text so the typesetter knows where it is to be inserted, for instance, "Insert A, p. 4"; the intended location is noted in the margin of the proof: "Insert A (attached)."

20 Posters

One of the chief attractions of scientific meetings is the presentations of papers, where the scientist can learn about the very latest research techniques and findings before they are published (which takes months), and have a person-to-person discussion with the attending authors. Years ago every conference attendee was allowed to present one or more papers in person. To accommodate all the speakers, several sessions would be held concurrently in separate conference rooms.

Eventually, participation at scientific meetings increased to the point where the number of speakers had to be restricted. For a time, societies appointed some of their members to serve on juries, selecting the papers to be presented orally and the rest as posters. This arrangement did not work because many authors felt slighted when their papers were chosen to be posters. Today only those who have received invitations to speak may give plenary lectures and symposia papers, but anyone who registers at the conference may present a paper in poster format. As there are very few speaking invitations issued, poster presentations are the norm. A famous scientist is as likely to have a poster as any other researcher.

So far this new arrangement appears to work well, but there are also disadvantages to the poster format. For one, a poster does not make use of your public-speaking skills. And if you are unskilled, a poster presentation will not give you the practice needed to improve. The ability to speak before an audience is important training for graduate students and postdoctoral fellows.

Today societies such as the *Federation of American Societies for Experimental Biology* (FASEB) can no longer hold their annual meeting with

This chapter is based in part on a lecture by Mary Helen Briscoe at the Cardiovascular Research Institute, UCSF. Presented here with the kind permission of Ms. Briscoe.

all their member societies present—the number of participants would easily exceed 20,000. Even special symposia such as the *Gordon Conferences* have over 100 participants for each conference. With so many posters there is considerable competition for the reader's time and attention.

Fig.15. Two types of posters. Illustration by M. H. Briscoe (private communication).

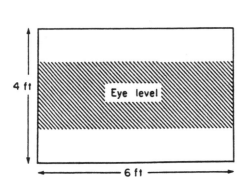
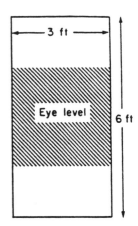

Fig. 16. A poster layout with the arrows showing the pathway. Illustration by M. H. Briscoe (private communication).

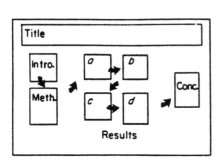
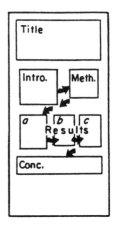

The posters are displayed at eye level so that the attendees can glance over them while walking. Most papers get very brief consideration under these circumstances, even though they have been segregated into many sessions by subject.

Your challenge is to attract the reader's attention long enough to communicate your findings. The essence of visual communication is to combine visual simplicity with concise and lucid text and an attractive layout. Basically, a poster is a large sign displaying a message or, in this case, a scientific paper. It usually measures 6 × 4 ft or 1.8 × 1.2 m. A rectangular poster can be oriented either horizontally or vertically (Figure 15), and a layout for smooth flow arranged accordingly (Figure 16).

There are many successful poster designs, but several principles apply to all. (For more examples of poster styles, see the critically acclaimed *A Researcher's Guide to Scientific and Medical Illustrations* by Mary Ellen Briscoe, 1990.)

1. Do not overload the poster with illustrations or text.

Many authors try to use up all the space on the poster, which is neither necessary nor desirable. Dense text is intimidating and difficult to read. When faced with a cluttered poster, many people will simply stop reading and move on to a less daunting display. Use white space to break up the text, which will make the poster more inviting and much easier to read. Compare the following posters:

$$E = mc^2$$

> Professor Albert Einstein is perhaps best known for his theory of relativity, which states that the amount of energy contained in an atom is equal to the mass of the atom multiplied by the square of the speed of light. Or, as most people have come to know it, $E = mc^2$. How did Einstein come up with this theory? Some say that the equation was the result of inspired insight, not hard work.

Less is more: Which poster would you rather read? Naturally, the first poster is more likely to be read.

2. Keep the text short and explicit.

A concise statement is communicated more effectively than lengthy prose. Remember that the "window of opportunity" is open for only a few seconds, during which the reader must look at the poster, understand the concept, and respond with interest. Only then will he read the entire poster and ask the attending author for more information.

The abstract is usually too long or complex to be of use in a poster. Some conferences require its inclusion, though, even if it is distributed in the meeting's *Abstracts*. It is really a waste of the poster space, and you might choose therefore to deemphasize it in your design.

3. Maximize legibility.

Expect that the reader will be standing about 3 ft or 1 m from the poster; both text and illustrations should be clear at this distance. To attract the reader, the letters in the title are even larger than those in the text; they should be at least 1 in or 2.5 cm high. Authors and affiliations should be in smaller type than the title.

If your subject is popular, the text should be even larger to accommodate a small crowd. Type is measured in points, one point being about 1/72 of an inch. Experiment with different typesizes and test them for legibility: can they be read without eyestrain? If necessary, delete some text to accommodate the typesize; legibility is more important than quantity.

Some authors practice calligraphy and prefer their own handwriting to printed posters, but these are often difficult to read. I believe that audiences are entitled to legible exhibits, and it is incumbent on the author to provide a computer-generated poster.

4. *Choose typestyles carefully.*

Do not try to incorporate every feature of your computer software. Bolding, graphics, varied fonts, etc., can add interest and emphasis if used sparingly, but it is a mistake to combine several styles. When too many styles are combined in one poster, the effect is what graphic designers describe as "a dog's breakfast." The following combination

The Effect of Sodium on HIGH BLOOD PRESSURE
A. B. Carter, University of California
San Francisco, California 94143

is really a distracting jumble. Even a single style, such as bolding, should not be applied to large portions of text or it will lose its effectiveness.

Some people set the text in uppercase for emphasis and because it allows the use of a smaller typesize. Be aware that uppercase letters take up more horizontal space and make the lines look crowded. Compare, for example,

A LOW-SODIUM DIET IS RECOMMENDED FOR ALL HIGH-RISK PATIENTS. CONSULT YOUR PHYSICIAN BEFORE MAKING ANY CHANGES TO DIET OR ACTIVITY.

versus

> A low-sodium diet is recommended for all
> high-risk patients. Consult your physician
> before making any changes to diet or activity.

The type in the first poster is 10% smaller than the type in the second poster, yet it takes up more lines because of the wider uppercase letters.

Consider also that the brain recognizes a word by its shape as well as by its letters, but all words printed in uppercase letters are rectangular and harder to recognize. Thus, mixed-case is easier to read than uppercase, which should only be used for short text such as titles and subheadings. Use mixed-case for the main text, even if a larger typesize is needed for legibility.

5. Clearly separate the text.

Text that is not visually separated is difficult to grasp, as illustrated by the following chain of sentences:

> The CD spectra of the two lectins showed a broad minimum around 200
> nm and a maximum above 190 nm. However, one spectrum (VVLM)
> was red-shifted by 5 nm from the other (VVLG). These findings
> suggest that both proteins are rich in β-sheets with little or no α-
> helices.

This long paragraph full of technical terms is too much to absorb at once. Judicious use of white space can break down unwieldy text into manageable sections.

> The CD spectra of the two lectins showed a broad
> minimum around 200 nm and a maximum above 190 nm.

However, one spectrum (VVLM) was red-shifted by 5 nm from the other (VVLG).

These findings suggest that both proteins are rich in β-sheets with little or no α-helices.

Also note the shorter line length, which is easier to read.

6. *Put the most important information at eye level.*

The most important information should be positioned at eye level (Figure 15), where it will receive the most attention. "Most important" information is not necessarily the findings; it could be another aspect of the project such as the research question, methodology, etc., as long as it holds the reader's interest.

7. *Use portrait orientation.*

"Portrait" or horizontal text is easier to read. Text in a journal can be rotated 90° to save space, without much inconvenience to the reader (Figure 17, top). Unlike a publication, however, fixed displays such as posters and slides cannot be rotated. People do not want to tilt their heads sideways to read, so orient all text horizontally (Figure 17, bottom).

8. *Arrange the poster components for continuous, smooth flow.*

Long lines of text are difficult to read because the eye must jump from the end of one line to the beginning of the next. The poster should be formatted into columns so that the lines are short enough to be read easily. Information should flow from one column to the next, left to right, as illustrated by the arrows in Figure 16. Note also that text for posters is often left-justified (ragged right margin) to maintain consistent wordspacing and letterspacing.

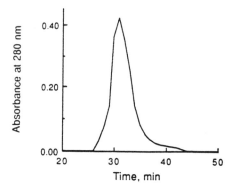

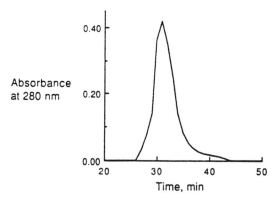

Fig. 17. Fast pressure liquid chromatography of a protein solution eluated with a linear gradient of increasing NaCl concentrations. *Top,* vertical lettering on the ordinate; *bottom,* horizontal lettering on the ordinate. Note that the horizontal lettering is easier to read than the vertical. Reprinted, with permission, from Tao, Shen, and Yang (1993) *J. Protein Chem.* **12**, 387–391. © 1993 by Plenum Publishing Corp., New York, NY.

21 Preparation of an Oral Presentation

Every researcher should be able to prepare and deliver good oral presentations. Some universities, recognizing the value of public-speaking skills, encourage graduate students to take a course in which they create visual aids and incorporate them into lectures. As there is no substitute for practice, you should take advantage of opportunities to give presentations in the local scientific community.

Public speaking is largely a matter of personal style. A good speaker can make a complex subject such as nuclear physics understandable to the audience, yet there are distinguished scientists, including some Nobel laureates, who simply cannot get their ideas across in a lecture.

Lectures are usually accompanied by visual aids, most often projections of slides or transparencies. Every illustration should be designed with the medium in mind. If the lecture is based on a paper intended for publication, the author may wish to use the same illustrations for both purposes. Unfortunately, illustrations suitable for publication are often too crowded for slides. For example, a slide of a DNA sequence with several hundred one-letter abbreviations cannot be read beyond the first several rows of the audience, nor is there time to observe more than one or two salient features. This kind of slide merely indicates that the authors have spent considerable time identifying the nucleotide residues.

The following guidelines are by no means rigid and should be modified at your own discretion.

1. Use large lettering for projections.

Slides are made from 35 mm film set into a 2 × 2 inch cardboard frame. Lettering in slides must be large enough to be legible from the last row of the audience. This task is made somewhat simpler when television monitors are strategically located in a huge auditorium. Use the expected attendance and room configuration to estimate what typesize to use.

For its national meetings, the American Chemical Society advises authors to use a template (2½ × 3¾ inches) for typewritten copy on 35 mm slides. It allows up to nine lines of text, double-spaced and wide enough for 54 elite or 45 pica characters. For a projected image 5 ft (1.5 m) high, mixed upper- and lowercase elite characters will be legible up to 40 ft (12 m) away; pica characters to 60 ft (18 m). The use of all uppercase letters will extend the legibility distance only slightly, and renders the text less intelligible (see the discussion on posters in chapter 19).

An illustration can be photocopied onto an 8½" × 11" transparency for use in an overhead projector. Again, any lettering must be large enough to be read from the last row, and if the original was handwritten, be sure that it is legible.

2. Include a heading on each illustration.

Each slide needs a heading or title to identify the information being displayed. Do not include discussion or interpretation, which belong in the lecture.

3. Limit each slide to one main idea.

4. Keep slides simple.

It is far better to present a sequence of simple slides than a single complicated one.

5. Number the slides in order.

Arrange the slides in display order, then use a pencil to number each slide carefully on its cardboard frame. Be sure to mark the unprinted side of each frame at the edge that should be uppermost when projected, so that the pictures will be oriented properly.

6. Make duplicates of slides that are used more than once.

Shuffling through slides during a presentation shows poor organization and is an imposition on the audience. Make copies of slides that are used more than once, so that you begin your lecture with all the slides in order.

7. If possible, use graphs rather than tables.

Graphs are usually more interesting and easier to interpret than rows of data presented in a table. Unlike graphs in publications, sideways text should not be used. Orient the ordinate label and other text horizontally so that the audience can read it without tilting their heads to the side.

8. Limit curves to a maximum of three or four per figure.

9. Limit columns to a maximum of four per table or seven per bar chart.

10. Limit text to a maximum of five to seven lines per slide and six or seven average-length words per line.

11. Leave some space between lines.

12. Use a pointer.

Often you will need to discuss a specific feature of the illustration. It is easier for the audience if you point directly to the feature, rather than making a statement such as "Note the data point just above the X-intercept." Pointers are available as sticks or light beams. The light beams are popular because you can point to the projection without leaving the podium. A pointing stick requires walking up to the illustration, which can be inconvenient, particularly if the microphone is stationary. Consider, though, that physical movement such as walking, pointing, and gesturing can make your lecture more dynamic.

13. Remember to turn on the room light at intervals.

Prolonged darkness can tire the audience, so turn on the room light after completing all the slides for a topic.

14. Do not read your lecture to the audience.

Reading from a script will drain the life out of your presentation. The most successful lecturers speak naturally and maintain eye contact, as though in conversation with the audience. Write a brief outline (3 × 5 inch note cards are good for this) of points to mention and when to display exhibits. Write the outline in large letters with space between items so you can find your place quickly. During your lecture, refer to the cards for the next topic, look away from the cards, and address the audience directly. Talk to different sections of the audience to involve more people in your lecture.

I vividly recall my first oral presentation at a national meeting of the American Chemical Society. I was well prepared, so much so that my lecture took less than the fifteen minutes allowed for each paper. However, I was nervous and simply read the entire lecture from my typed notes. Although a well-known scientist later assured me that my work was well done, I considered it a failure in that it completely lacked spontaneity. Since then, I still prepare for a lecture and may even type up the entire speech, but during the presentation I rely on only the outline.

Appendices

Appendix A. IUPAC Symbols and Terminology for Physicochemical Quantities and Units

Authors are advised to adopt the conventions presented by the International Union of Pure and Applied Chemistry (IUPAC) in the *IUPAC Manual of Symbols and Terminology for Physicochemical Quantities and Units* (Butterworths, London, 1970). The Manual is also available in *National Bureau of Standards Special Publication 330* (U.S. Government Printing Office, Washington, DC, 1981) and *Pure and Applied Chemistry* (1970), **21**, 3-44.

IUPAC codified scientific terminology into a universally-accepted standard that has been largely adopted by the scientific community, although some older scientific terms still persist. One example is sodium dodecyl sulfate, which is almost universally abbreviated as *SDS*. IUPAC objects to its use of *S* to represent both sodium and sulfate, and advocates using *NaDodSO$_4$* instead. This abbreviation is somewhat unwieldy and cannot be expected to supplant SDS, which is both convenient and securely established in the lexicon. Thus, some journals allow the use of either abbreviation. Remember, though, that it is never incorrect to use IUPAC terms.

SI UNITS

IUPAC advocates the International System of Units (*Le Système International d'Unités* in French), abbreviated *SI*. IUPAC defines a physical quantity, represented by a symbol, as its unit multiplied by a (dimensionless) number:

$$\textit{Physical quantity} = \text{Numerical value} \times \text{Unit}$$

For example, the pathlength of a one-millimeter optical cell is expressed as $l = 1$ mm.

All symbols are given in italic type; symbols for vectors (quantities with magnitude and direction components) are bolded as well. Abbreviations for SI units are given in lowercase Roman (i.e., upright) type with no period, and the same abbreviation is used for both singular and plural. Thus, the length of five centimeters is written as 5 cm, not 5 cm. or 5 cms.

The subscript in a symbol is not italicized unless it is itself a symbol. For instance, the concentration of substance B is written as C_B, but heat capacity at constant pressure is C_p because p is a symbol.

SI has five fundamental elements or "base units" (Table A-1.1). It has additional units that derive from the base units (Table A-1.2).

Some IUPAC recommendations:

1. In the base quantities, *n, m* and *l* are symbols for quantities, not for numbers. It is therefore wrong to call *n* the "number of moles."

2. The term *specific* means "divided by mass" and the term *molar* "divided by amount of substance." Thus,

$$\text{specific volume} \quad v_{sp} = V/m$$

and

$$\text{molar volume} \quad V_m = V/n$$

Table A-1.1 Base physical quantities

Base physical quantity	Symbol	Quantity	Unit
length	l	meter	m
mass	m	kilogram	kg
time	t	second	s
thermodynamic temperature	T	Kelvin	K (not °K)
amount of substance	n	mole	mol

Table A-1.2 Some common units of measurement

Quantity	Unit	Quantity	Unit
Length		Electricity and magnetism	
centimeter	cm	ampere	A
millimeter	mm	volt	V
micrometer	μm(not λ)	ohm	Ω
nanometer	nm (not mμ)	siemens (mho)	S
angstrom (0.1 nm)	Å	gauss	G
Mass		Energy	
gram	g	joule	J
milligram	mg	calorie	cal
microgram	μg	kilocalorie	kcal
Time		Radioactivity	
minute	min	becquerel[†]	Bq
hour	h	curie[†]	Ci
day	d	counts per minute	cpm
Temperature		disintegrations per minute	dpm
degree Celsius	°C (deg C)	Other units	
Volume		cycles per minute (hertz)	Hz
milliliter[*]	ml	dalton	Da
microliter	μl (not λ)	degree (angle)	deg
square centimeter	cm^2	parts per million	ppm
cubic centimeter	cm^3	revolutions per minute	rpm
Concentration		svedberg (10^{-13} s)	S
molar (mol/liter)	M	wave number	cm^{-1}
millimolar	mM		

Note: The five base units are listed in Table A-1.1.
[*]Some journals specify that *liter* should not be abbreviated unless used with a prefix, whereas others such as *Biochemistry* use uppercase L (e.g., mL) instead of lowercase *l*, to avoid confusion with the numeral 1.
[†]Ci = 3.7 x 10^{10} Bq (or 37 GBq; see Table A-2); 1 Bq = 60 dpm.

3. Products and quotients: Use parentheses to avoid ambiguity when operations are combined. For example,

ab
a·b
a×b

are all acceptable ways to express the product of a and b, and

(a/b)/c
a/(bc)
a/(b/c)
ac/b

are unambiguous, but

a/b/c

could cause confusion and should be avoided.

4. IUPAC recommends relative molar mass for molecular weight M_r (dimensionless), a term not yet widely used. Thus, the unit dalton (Da) for molecular weight is dimensionless. On the other hand, the unit of molar mass is g/mol. Confusion may arise when the following examples mean the same thing:

the relative molar mass of X is 10^6
the molecular weight of X is 10^6
the molecular mass of X is 10^6 daltons
the molar mass of X is 10^6 g
the 10^6-dalton X

Note also that the expression "273 K" refers to the absolute temperature on the Kelvin scale, i.e., the freezing point of water. It is not the molecular weight 273,000.

5. The old definition of *liter* (1.000,028 dm^3) is rescinded. Thus, 1 *l* (or *L*) is now 1 dm^3 or 1,000 cm^3. The term *milliliter* (ml or mL), is still used and is just 1 cm^3. Note that the unit for liter is given in italics instead of Roman type to avoid confusion with the numeral 1.

6. In chemistry, 0.1 mol dm^{-3} is defined as 0.1 M. In mks units, a 0.1 kg/m^3 solution may be incomprehensible to us chemists, but it is really equivalent to 1 mg/ml.

7. For spectrophotometric data:

$$A = -\log_{10} T$$

and

$$A = \epsilon l C$$

where A is the absorbance (dimensionless), T the transmittance (also dimensionless and equal to I/I_o or the ratio of exit to incident intensity), ϵ the molar absorption coefficient, C the concentration of the absorbing substance in mol/dm^3, and l the length of optical path in cm. Therefore, the dimension of ϵ is $M^{-1}cm^{-1}$.

The term *extinction coefficient* is now reserved for diffusion of radiation. The term *molar absorptivity* should be avoided because absorptivity now means absorbance per unit length. The term *optical density* is now used for light transmission through turbid suspension, as distinguished from absorbance for light absorption by solution.

8. Avoid applying superscripts to subscripts or subscripts to superscripts. For example,

$$\Delta H_{25\,°C}$$

should be written as

$$\Delta H \text{ at } 25\,°C$$

SI PREFIXES

In SI units, the use of prefixes is preferred to power-of-ten notation (Table A-2); additional units are created by prefixing the term for a power of ten to a base unit. For example, the kilometer is a unit equal to 10^3 meters. Abbreviations for these derived units are formed by joining the abbreviations for the prefix and the base unit. Thus, km is the abbreviation for kilometer.

Note that the base unit for mass (kilogram) already has a prefix (kilo). Derived units for mass are formed by applying the prefixes to gram to form decigram (dg), megagram (Mg), and so on.

Table A-2   Prefixes to names of units

Factor	Prefix	Abbreviation	Factor	Prefix	Abbreviation
10^{-1}	deci	d	10^{1}	deca	da
10^{-2}	centi	c	10^{2}	hecto	h
10^{-3}	milli	m	10^{3}	kilo	k
10^{-6}	micro	μ	10^{6}	mega	M
10^{-9}	nano	n	10^{9}	giga	G
10^{-12}	pico	p	10^{12}	tera	T
10^{-15}	femto	f	10^{15}	peta	P
10^{-18}	atto	a	10^{18}	exa	E

FUNDAMENTAL CONSTANTS

Refer to Table A-3.

Table A-3   Some fundamental constants

Quantity	Symbol	Value
Acceleration of gravity	g	9.80665 m·s^{-2}
Avogadro's constant	L, N_A	$6.022045 \times 10^{23} \text{ mol}^{-1}$
Boltzmann's constant	$k = R/T$	$1.380662 \times 10^{23} \text{ J·K}^{-1}$
Gas constant	R	$8.31141 \text{ J·K}^{-1}\text{·mol}^{-1}$
Planck's constant	h	$6.626176 \times 10^{-34} \text{ J·Hz}^{-1}$
	$h/2\pi$	$1.10545887 \times 10^{-33} \text{ J·s}$
Speed of light in vacuum	c	$2.99792458 \times 108 \text{ m·s}^{-1}$

COMMON PHYSICAL AND CHEMICAL QUANTITIES

Refer to Table A-4.

Table A-4 Some common physical and chemical quantities

Quantity	Symbol
absorbance*	A
density	r
diffusion coefficient	D
equilibrium constant	K
maximum velocity	V_{max}
Michaelis constant	K_m
optical density	OD
partial specific volume	ν
rate constant	k
relative molecular mass (*or* molecular weight)	M_r
retardation factor	R_f
sedimentation coefficient	S
specific rotation	$[\alpha]$
Thermodynamic quantities:	
Gibbs free energy change	ΔG
entropy change	ΔS
enthalpy change	ΔH
standard deviation	SD
standard error	SE
standard error of mean	SEM

*Use absorbance A for light absorption by solution, and optical density OD for light transmission through turbid suspensions. (OD is not a symbol in SI and is therefore set in Roman type.)

Appendix B. Some Standard Abbreviations and Symbols

Standard abbreviations for common substances need not be defined in ▪e text. These include the symbols for the chemical elements, 3- and 1-▪tter codes for the amino acids, 3-letter codes for carbohydrates, lipids, and ▪ucleotides, and 2-letter codes for chemical radicals.

Tables B-1 through B-7 list some standard symbols and abbreviations ▪sed in biochemistry and molecular biology.

Symbols for Pyrimidine and Purine Bases and Nucleosides

Refer to Table B-1.

Symbols for Nucleotides

Refer to Table B-2.

Symbols for Amino Acids

Refer to Table B-3.

The Standard Genetic Code

Refer to Table B-4.

Isotopically-Labeled Compounds

An isotope is represented by the symbol for element prefixed with the atomic weight in superscript, and the whole enclosed in brackets. Thus, the symbol for the isotope carbon-14 is $[^{14}C]$. The presence of one or more isotopes in a compound is denoted by prefixing the isotope symbol(s) to the regular name or formula of the compound.

$[^{14}C]$urea

$[\alpha\text{-}^{32}P]ATP$

$[^{32}P]CMP$ (not $CM^{32}P$)

$[^{2}H]CH_4$

$[3\text{-}^{14}C, 2\text{-}, 3\text{-}^{3}H, \,^{14}N]$serine

L-[methyl-^{14}C]methionine

^{131}I-labeled albumin (but not $[^{131}I]$albumin, because native albumin does not contain iodine; however, ^{131}I-albumin and $[^{131}I]$iodoalbumin are both acceptable)

$[U\text{-}^{14}C]$glucose (where the ^{14}C is uniformly distributed)

$[G\text{-}^{14}C]$glucose (where ^{14}C is distributed among all six positions, but not necessarily uniformly)

Table B-1 Symbols for Pyrimidine and Purine Bases and Nucleosides

	3-letter	1-letter
Adenine	Ade	
Cytosine	Cyt	
Guanine	Gua	
Thymine	Thy	
Uracil	Ura	
Xanthine	Xan	
Adenosine	Ado	A
Cytidine	Cyd	C
Guanosine	Guo	G
Inosine	Ino	I
Pseudouridine*	Ψrd	Ψ or Q
Ribosylthymine	Thd	T
Thymidine	dThd	dT
Uridine	Urd	U
Xanthosine	Xao	X
Phosphoric residue†	-*P*	p or -

*The one-letter symbol Q is used for computer work.

†In the 3-letter system for the nucleosides, a capital italic *P* is used for the phosphoric residue, and in the 1-letter system a lowercase p or - is used. The 1-letter symbols are used for the nucleoside residues in the sequence.

Table B-2   Symbols for Nucleotides

Abbreviations	Name
AMP, ADP, and ATP	Adenosine 5'-mono-, di-, and triphosphates
dAMP, etc.	DeoxyAMP, etc.
cAMP, etc.	CyclicAMP, etc.
CMP, CDP, and CTP	Cytidine 5'-mono-, di-, and triphosphates
dCMP	DeoxyCMP, etc.
dTMP, dTDP, and dTTP	Thymidine 5'-mono-, di-, and triphosphates
GMP, GDP, and GTP	Guanosine 5'-mono-, di-, and triphosphates
dGMP, etc.	DeoxyGMP, etc.
IMP, IDP, and ITP	Inosine 5'-mono-, di-, and triphosphates
TMP, TDP, and TTP	Ribosylthymidine 5'-mono-, di-, and triphosphates
UMP, UDP, and UTP	Uridine 5'-mono-, di-, and triphosphates
dUMP, etc.	DeoxyUMP, etc.
Pi	Inorganic phosphate
PPi	Inorganic pyrophosphate
DNA	Deoxyribonucleic acid or deoxyribonucleate
RNA	Ribonucleic acid or ribonucleate

Table B-3   Symbols for Amino Acids

	3-letter	1-letter		3-letter	1-letter
Alanine	Ala	A	Methionine	Met	M
Asparagine or aspartic acid	Asx	B	Asparagine	Asn	N
Cysteine	Cys*	C	Proline	Pro§	P
Aspartic acid	Asp	D	Glutamine	Gln	Q
Glutamic acid	Glu	E	Arginine	Arg	R
Phenylalanine	Phe	F	Serine	Ser	S
Glycine	Gly	G	Threonine	Thr	T
Histidine	His	H	Valine	Val	V
Isoleucine	Ile	I	Tryptophan	Trp	W
Lysine	Lys†,‡	K	Tyrosine	Tyr	Y
Leucine	Leu	L	Glutamine or glutamic acid	Glx	Z

*Cys-, half-cystine.
†Hyl, hydroxylysine.
‡Orn, ornithine (a homologue of Lys with one less methylene group in the side chain).
§Hyp, hydroxyproline.

Table B-4   Symbols for triple-nucleotide codes and 3-letter amino acids

UUU	Phe	UCU	Ser	UAU	Tyr	UGU	Cys
UUC	Phe	UCC	Ser	UAC	Tyr	UGC	Cys
UUA	Leu	UCA	Ser	UAA	stop	UGA	stop
UUG	Leu	UCG	Ser	UAG	stop	UGG	Trp
CUU	Leu	CCU	Pro	CAU	His	CGU	Arg
CUC	Leu	CCC	Pro	CAC	His	CGC	Arg
CUA	Leu	CCA	Pro	CAA	Gln	CGA	Arg
CUG	Leu	CCG	Pro	CAG	Gln	CGG	Arg
AUU	Ile	ACU	Thr	AAU	Asn	AGU	Ser
AUC	Ile	ACC	Thr	AAC	Asn	AGC	Ser
AUA	Ile	ACA	Thr	AAA	Lys	AGA	Arg
AUG	Met	ACG	Thr	AAG	Lys	AGG	Arg
GUU	Val	GCU	Ala	GAU	Asp	GGU	Gly
GUC	Val	GCC	Ala	GAC	Asp	GGC	Gly
GUA	Val	GCA	Ala	GAA	Glu	GGA	Gly
GUG	Val	GCG	Ala	GAG	Glu	GGG	Gly

Table B-5   Some specific nucleic acids

	Symbol
Complementary DNA, RNA	cDNA, cRNA
Messenger RNA	mRNA
Mitochondrial DNA, RNA	mtDNA, mtRNA
Ribosomal RNA	rRNA
Small cytoplasmic RNA	scRNA
Small nuclear RNA	snRNA
Transfer RNA	tRNA

Source: "Instructions to Authors" in the *Journal of Biological Chemistry.*